T0269256

DEVELOPING THE GLOBAL BIOECONOMY

DEVELOPING THE GLOBAL BIOECONOMY

TECHNICAL, MARKET, AND ENVIRONMENTAL LESSONS FROM BIOENERGY

Edited by

PATRICK LAMERS
Idaho National Laboratory, Idaho Falls, ID, USA

ERIN SEARCY
Idaho National Laboratory, Idaho Falls, ID, USA

J. RICHARD HESS
Idaho National Laboratory, Idaho Falls, ID, USA

HEINZ STICHNOTHE
Thünen Institute of Agricultural Technology, Braunschweig, Germany

AMSTERDAM • BOSTON • HEIDELBERG • LONDON
NEW YORK • OXFORD • PARIS • SAN DIEGO
SAN FRANCISCO • SINGAPORE • SYDNEY • TOKYO

ELSEVIER

Academic Press is an imprint of Elsevier

Academic Press is an imprint of Elsevier
125 London Wall, London EC2Y 5AS, UK
525 B Street, Suite 1800, San Diego, CA 92101-4495, USA
50 Hampshire Street, 5th Floor, Cambridge, MA 02139, USA
The Boulevard, Langford Lane, Kidlington, Oxford OX5 1GB, UK

Copyright © 2016 Elsevier Inc. All rights reserved.
Patrick Lamers, Erin Searcy and J. Richard Hess's contribution to the Work are in
public domain.

No part of this publication may be reproduced or transmitted in any form or by any
means, electronic or mechanical, including photocopying, recording, or any information
storage and retrieval system, without permission in writing from the publisher. Details on
how to seek permission, further information about the Publisher's permissions policies
and our arrangements with organizations such as the Copyright Clearance Center and the
Copyright Licensing Agency, can be found at our website: www.elsevier.com/permissions.

This book and the individual contributions contained in it are protected under copyright
by the Publisher (other than as may be noted herein).

Notices
Knowledge and best practice in this field are constantly changing. As new research and
experience broaden our understanding, changes in research methods, professional practices,
or medical treatment may become necessary.

Practitioners and researchers must always rely on their own experience and knowledge
in evaluating and using any information, methods, compounds, or experiments described
herein. In using such information or methods they should be mindful of their own safety
and the safety of others, including parties for whom they have a professional responsibility.

To the fullest extent of the law, neither the Publisher nor the authors, contributors, or
editors, assume any liability for any injury and/or damage to persons or property as a
matter of products liability, negligence or otherwise, or from any use or operation of any
methods, products, instructions, or ideas contained in the material herein.

British Library Cataloguing-in-Publication Data
A catalogue record for this book is available from the British Library.

Library of Congress Cataloging-in-Publication Data
A catalog record for this book is available from the Library of Congress.

ISBN: 978-0-12-805165-8

For Information on all Academic Press publications
visit our website at http://www.elsevier.com/

Working together
to grow libraries in
developing countries

www.elsevier.com • www.bookaid.org

Publisher: Joe Hayton
Acquisition Editor: Raquel Zanol
Editorial Project Manager Intern: Ana Claudia Abad Garcia
Production Project Manager: Kiruthika Govindaraju
Designer: Greg Harris

Typeset by MPS Limited, Chennai, India

DEDICATION

We dedicate this book to Jody Michelle Endres (1970–2015), former Assistant Professor of Law in the Department of Natural Resources and Environmental Sciences at the University of Illinois at Urbana-Champaign, United States. A devoted scholar and passionate advocate for the operationalization of sustainability standards, land use classifications, and the role of law in bioenergy modeling, her additions to the research field were outstanding. She will be missed.

DISCLAIMER

Part of the information presented in this book was prepared as an account of work sponsored by an agency of the US Government. Neither the US Government nor any agency thereof, nor any of their employees, makes any warranty, express or implied, or assumes any legal liability or responsibility for the accuracy, completeness, or usefulness of any information, apparatus, product, or process disclosed, or represents that its use would not infringe privately owned rights. References herein to any specific commercial product, process, or service by trade name, trademark, manufacturer, or otherwise, does not necessarily constitute or imply its endorsement, recommendation, or favoring by the US Government or any agency thereof. The views and opinions of authors expressed herein do not necessarily state or reflect those of the US Government or any agency thereof.

FINANCIAL AND COMPETING INTERESTS DISCLOSURE

The work performed by the employees of the Idaho National Laboratory was supported by the US Department of Energy under Department of Energy Idaho Operations Office Contract No. DE-AC07-05ID14517. The US Government retains and the publisher, by accepting this work for publication, acknowledges that the US Government retains a nonexclusive, paid-up, irrevocable, worldwide license to publish or reproduce the published form of this book, or allow others to do so, for US Government purposes. The authors have no other relevant affiliations or financial involvement with any organization or entity with a financial interest in or financial conflict with the subject matter or materials discussed in the manuscript apart from those disclosed. No writing assistance was utilized in the production of this manuscript.

CONTENTS

LIST OF FIGURES

LIST OF CONTRIBUTORS

B. Batidzirai
Energy Research Centre, University of Cape Town, Cape Town, South Africa

M. Beermann
Joanneum Research Forschungsgesellschaft mbH, Graz, Austria

I. de Bari
Division of Bioenergy, Biorefinery and Green Chemistry, ENEA Centro Ricerche Trisaia, Policoro, Italy

M. Deutmeyer
Green Carbon Group, Hamburg, Germany

R. Diaz-Chavez
Centre for Environmental Policy, Imperial College, London, United Kingdom

J. Heinimö
Mikkeli Development Miksei Ltd, Mikkeli, Finland

B. Hektor
Svebio, Stockholm, Sweden

J.R. Hess
Idaho National Laboratory, Idaho Falls, ID, United States

K. Johnson
US Department of Energy, Bioenergy Technologies Office, Golden, CO, United States

G. Jungmeier
Joanneum Research Forschungsgesellschaft mbH, Graz, Austria

M. Junginger
Copernicus Institute, Utrecht University, Utrecht, The Netherlands

M. Klemm
German Biomass Research Center (DBFZ), Leipzig, Germany

P. Lamers
Idaho National Laboratory, Idaho Falls, ID, United States

D. Meier
Thünen Institute of Wood Research, Hamburg, Germany

O. Olsson
Stockholm Environment Institute, Stockholm, Sweden

T. Ranta
University of Lappeenranta, Lappeenranta, Finland

F. Schipfer
EEG, Vienna University of Technology, Vienna, Austria

E. Searcy
Idaho National Laboratory, Idaho Falls, ID, United States

H. Stichnothe
Thünen Institute of Agricultural Technology, Braunschweig, Germany

H. Storz
Thünen Institute of Agricultural Technology, Braunschweig, Germany

S. Thomas
US Department of Energy, Bioenergy Technologies Office, Golden, CO, United States

D. Thrän
German Biomass Research Center (DBFZ); Department of Bioenergy, Helmholtz Center for Environmental Research (UFZ), Leipzig, Germany

E. Trømborg
Norwegian University of Life Sciences, Ås, Akershus, Norway

M. Wild
Wild & Partner, Vienna, Austria

BIOGRAPHY

Dr Patrick Lamers

Patrick Lamers is a Systems Analyst with the Idaho National Laboratory, stationed at the National Bioenergy Center in Golden, Colorado. His work on feedstock mobilization for the US Department of Energy's Bioenergy Technologies Office supports the deployment and scale-up of the US advanced biofuels industry. Patrick's academic degrees include a MSc from Karlsruhe Institute of Technology, Germany, a MSc from Lund University, Sweden, and a PhD in Energy and Resources from Utrecht University, the Netherlands. He has been a senior researcher and consultant across North America and Europe for well over 10 years, performing work for the private sector, governmental, as well as nongovernmental agencies. He has published extensively in the area of global biomass markets and is engaged in multiple international working groups such as the IEA Bioenergy.

Dr Erin Searcy

Erin Searcy is currently leading the Bioenergy Analysis Platform at the Idaho National Laboratory (INL). She originally joined INL in 2008 and has worked on a variety of biomass feedstock logistics projects since, primarily as a techno-economic analyst. Between 2012 and 2015, Erin was stationed at the US Department of Energy in Washington, DC, supporting the Bioenergy Technologies Office. Her academic degrees include a BSc and MSc in Engineering, as well as a PhD in Mechanical Engineering from the University of Alberta, Canada. Prior to joining INL, Erin had worked as an Environmental Engineering consultant and acted as a sessional professor in the Faculty of Engineering at the University of Alberta, Canada.

Dr J. Richard Hess

J. Richard Hess is the Director for the Idaho National Laboratory (INL) Energy Systems and Technologies Division, which addresses critical national energy challenges in biofuels/bioenergy, renewable electrical systems/grid, and hybrid renewable-nuclear systems. He led the development of a biomass feedstock preprocessing and logistics program at INL and continues to serve as the Laboratory Relationship Manager for that program. This program focuses on the cost-effective use of lignocellulosic biomass crops and residues in biorefining operations, including biomass harvesting, handling, storage and transportation; and preprocessing biomass into suitable industrial-grade bioenergy commodities through enhanced feedstock formulation, densification, and packaging for transportation. He also managed the design and construction of one of DOE's five biomass demonstration units, which in this case, was the Feedstock Process Demonstration Unit. Richard holds a Doctorate in Plant Science from Utah State University, and MSc and BSc Degrees in Botany from Brigham Young University. Following Graduate School, Richard served as an Agriculture Congressional Science Fellow in the Washington, DC Office of Senator Thomas Daschel. In this role, he worked on several national agricultural issues—including new and industrial uses of agricultural products, federal grain inspection standards, plant variety protection, and other agricultural R&D policy issues.

Dr Heinz Stichnothe

Heinz holds a BSc in Chemical Engineering and a PhD in Chemistry. He is senior scientist at the Thünen-Institute of Agricultural Technology. Heinz acts as evaluator for EU-BBI-JU, is vice-chair of the SETAC Europe LCA steering committee, was a member of the German delegation for developing ISO13065, and was involved in drafting the German Biorefinery Roadmap. He is the national representative for IEA Bioenergy Task 42 (Biorefining).

Heinz's research interests are in the area of engineering for sustainable development, which includes optimization of biotechnological and chemo-catalytic conversion processes of agricultural biomass and residues. He uses sustainability assessment, lifecycle assessment, and carbon footprint analysis of biobased systems and products in order to steer the development of biomass conversion processes in the most promising direction already at an early development stage. His ultimate goal is to foster strategies for the sustainable use of biomass for nonfood applications by providing advice to process developers and policymakers.

PREFACE

IEA BIOENERGY

IEA Bioenergy is an organization set up in 1978 by the International Energy Agency (IEA) with the aim of improving cooperation and information exchange between countries that have national programs in bioenergy research, development and deployment. As an international collaboration in bioenergy, IEA Bioenergy's vision is to achieve a substantial bioenergy contribution to future global energy demands by accelerating the production and use of environmentally sound, socially accepted, and cost-competitive bioenergy on a sustainable basis, thus providing increased security of supply whilst reducing greenhouse gas emissions from energy use. The work of IEA Bioenergy is structured in a number of Tasks, which have well defined objectives, budgets, and time frames. More information on IEA Bioenergy is available at: http://www.ieabio energy.com/.

TASK 34

This task focuses on facilitating commercialization of biomass pyrolysis, and particularly fast pyrolysis to maximize liquid product yield and production of renewable fuel oil and transportation fuels by contributing to the resolution of critical technical areas and disseminating relevant information particularly to industry and policymakers. The task contributes to bio-oil standardization activities and reviews bio-oil production techniques and bio-oil application. The task cooperates with application developers and equipment manufacturers to help them understand more about bio-oil and its properties and requirements. This close cooperation is considered the most effective way of identifying and promoting opportunities for bio-oil to make a significant contribution to renewable energy supplies. More information on Task 34 is available at: http://www.pyne.co.uk/.

TASK 40

The core objective of this task is to support the development of sustainable, international markets and international trade of biomass, recognizing the diversity in biomass resources and applications for bioenergy and bio-materials in the bioeconomy. Developing the sustainable and stable, international biomass market for energy and materials is a long-term process. It is particularly important to develop both supply and demand for biomass and energy carriers derived from biomass in a balanced way and avoid distortions and instability that can threaten investments in biomass production, infrastructure and conversion capacity. The task aims to provide a vital contribution to such (policy) decisions for market players, policymakers, and international bodies, as well as nongovernmental organizations. It aims to do so by providing high-quality information and analyses, such as this book. The focus on international biomass trade is a priority, although biomass will be increasingly utilized in the bioeconomy for new material purposes (eg, bioplastics) and thus find alternative uses before or in conjunction with the use for energy. More information on Task 40 is available at: http://www.bioenergytrade.org/.

TASK 42

This task aims at contributing to the development and implementation of sustainable biorefineries—as part of highly efficient and zero waste value chains—synergistically producing food and feed ingredients, bio-based chemicals and materials and bioenergy (fuels, power/heat) as a base for a global bioeconomy. The mission of Task 42 is to facilitate the commercialization and market deployment of environmentally sound, socially acceptable and cost-competitive biorefineries, and to advise policymakers and industrial decision-makers accordingly. Its strategy is to provide a platform for international collaboration and information exchange concerning biorefinery research, development, demonstration, and policies. More information on Task 42 is available at: http://www.iea-bioenergy.task42-biorefineries.com/en/ieabiorefinery.htm/.

ACKNOWLEDGMENTS

Our colleagues from the IEA Bioenergy, particularly Tasks 34, 40, and 42, provided valuable comments and suggestions to earlier drafts of this work. The INL employees also wish to thank the US Department of Energy for funding and supporting their work.

CHAPTER 1

Bioeconomy Strategies

J.R. Hess[1], P. Lamers[1], H. Stichnothe[2], M. Beermann[3] and G. Jungmeier[3]
[1]Idaho National Laboratory, Idaho Falls, ID, United States
[2]Thünen Institute of Agricultural Technology, Braunschweig, Germany
[3]Joanneum Research Forschungsgesellschaft mbH, Graz, Austria

Contents

Abstract

Facing a shortage of petrochemicals in the long term, biomass is expected to be the main future feedstock for chemicals, including liquid transportation fuels. Currently, biomass is mainly used for food, feed, and material purposes; only a small fraction is used in energy conversion (ie, heating/cooling, power, or transport fuels). The "bioeconomy" has been referred to as the set of economic activities that relate to the invention, development, production and use of biological products and processes. The transition from an economy based on fossil raw materials to a bioeconomy, obtaining its raw materials from renewable biological resources requires concerted efforts by international institutions, national governments, and industry sectors, and prompts for the development of bioeconomy policy strategies. However, there is still little understanding on how current markets will transition towards a national and essentially global bioeconomy. This joint analysis brings together expertise from three IEA Bioenergy subtasks: Task 34 on Pyrolysis, Task 40 on International Trade and Markets, and Task 42 on Biorefineries. The underlying hypothesis is that bioeconomy market developments can benefit from lessons learned and developments observed in bioenergy markets. The question is not only how the bioeconomy can be developed, but also how it can be developed sustainably in terms of economic and environmental concerns. The strength of bringing three IEA Bioenergy subtasks into this analysis is found in each task's area of expertise. Tasks 34 and 42 identify the types of biorefineries that are expected to be implemented and the types of feedstock that may be used. Task 40 provides complementary work including a historical analysis of the developments of biopower and biofuel markets, integration opportunities into existing supply chains, and the conditions that would need to be created and enhanced to achieve a biomass supply system supporting a global bioeconomy.

Developing the Global Bioeconomy.
DOI: http://dx.doi.org/10.1016/B978-0-12-805165-8.00001-X
© 2016 Elsevier Inc.
All rights reserved.
1

1.1 INTRODUCTION

Reducing and replacing the utilization of fossil resources is among the most critical challenges in transforming the current energy supply system and consumption patterns (IEA, 2014; IPCC, 2014). Although the exploration of unconventional fossil resources (shale gas, tar sand, etc.) has expanded the spectrum of exploitable *resources*, fossil resources remain finite and are not readily renewable. The observed increase in global mean surface temperature over the past decades is very likely due to anthropogenic greenhouse gas (GHG) emissions (IPCC, 2007, 2014). It is generally assumed that GHG-induced climate change can be mitigated by efficiency improvements, sequestration of CO_2, and by shifting from fossil primary energy resources to a variety of renewable resources (Trainer, 2010). Facing a shortage of petrochemicals in the long term, biomass is expected to be the main future feedstock for chemicals, including liquid transportation fuels (Langeveld et al., 2010). Currently, biomass is mainly used for food, feed, and material purposes; only a small fraction is used in energy conversion (ie, heating/cooling, power, or transport fuels).

The "bioeconomy" refers to the set of economic activities that relate to the invention, development, production, and use of biological products and processes (OECD, 2009). Within the context of this work, bioeconomy is defined as the economic, environmental and social activities associated with the production, harvest, transport, preprocessing, conversion, and use of biomass for biopower, bioproducts, and biofuels. Within a future bioeconomy, biomass will be used for the sustainable and synergetic production of food, feed, bioenergy (power, heat, and biofuels) and biobased products (chemicals and materials) (Bell et al., 2014; NEA, 2014). The main industrial sectors likely to be involved in the future bioeconomy are agriculture and forestry, and include their related processing industries (eg, food and feed, pulp and paper, etc.), plus chemicals and materials (Fig. 1.1). The production of biobased materials is not new. However, the majority of fuels, nitrogen fertilizers, organic chemicals, and polymers are still derived from fossil-based feedstock, predominantly oil and gas.

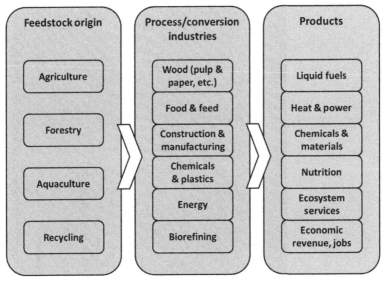

Figure 1.1 Overview of the bioeconomy as understood within the context of this work.

1.2 STATUS OF BIOECONOMY STRATEGIES IN IEA BIOENERGY MEMBER COUNTRIES

The transition from an economy based on fossil raw materials to a bioeconomy, obtaining its raw materials from renewable biological resources, requires concerted efforts by international institutions, national governments, and industry sectors, and prompts for the development of bioeconomy policy strategies. In 2012, the European Commission (EC) launched a new strategy on the bioeconomy (EC, 2012). In this communication, the EC states that Europe needs to radically change its approach to production, consumption, processing, storage, recycling, and disposal of biological resources, in order to cope with an increasing global population, rapid depletion of many resources, increasing environmental pressures, and climate change.

The current progress and priorities of such strategies worldwide are summarized in the following as result of a survey in the 22 member countries of the IEA Bioenergy Implementing Agreement (Fig. 1.2, Table 1.1). The survey (Beermann et al., 2014) is based on information and documents available up to September 2014.

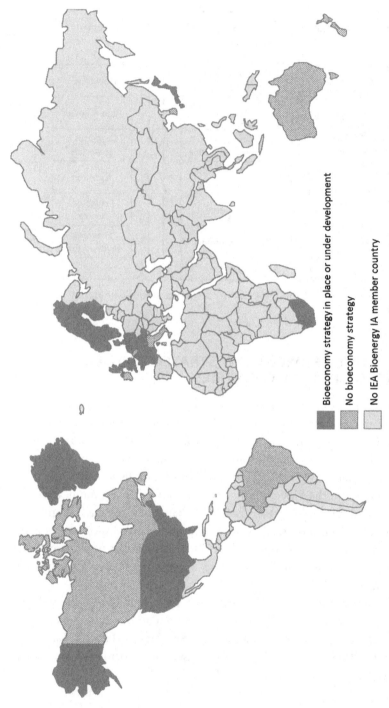

Figure 1.2 Bioeconomy strategies across IEA Bioenergy Implementing Agreement (IA) member countries as of September 2014 (Beermann et al., 2014).

Bioeconomy strategy in place or under development

No bioeconomy strategy

No IEA Bioenergy IA member country

Table 1.1 Bioeconomy strategies in the member countries of the IEA Bioenergy Implementing Agreement (status: September 2014) (Beermann et al., 2014)

		AU	AT	BE	BR	CA	HR	DK	FI	FR	DE	IT	JP	NL	NZ	NO	ZA	SE	CH	UK	US
Strategies	Governmental bioeconomy	x	(√)	x	x	x	x	(√)	√	(√)	√	x	(√)	√	x	(√)	√	√	(√)	(√)	(√)
	Industry	√	x	x	x	√	x	√	x	x	√	√	√	x	√	x	x	√	x	x	x
	Regional	√	x	x	√	√	x	x	x	x	√	x	√	x	x	x	x	√	x	x	x
	Policy advice	√	√	√	√	√	√	√	√	√	√	√	√	√	√	√	√	√	√	√	√
Strategy scope	Bioeconomy	x	√	√	√	√	√	√	√	√	√	√	x	√	√	x	x	√	√	x	x
	Biobased economy	√	√	√	√	√	√	√	√	√	x	√	√	√	√	√	√	√	√	√	x
	Biobased industries	√	√	√	√	√	√	√	x	√	x	√	√	x	√	√	√	x	√	√	√
Role of bioenergy	Priority	x	x	x	x	x	x	x	√	x	√	x	√	x	√	x	x	x	x	√	√
	Equal to other sectors	√	√	√	√	√	√	√	x	√	√	√	x	√	√	√	√	√	√	x	x
	Less importance	x	x	x	x	x	x	x	x	x	x	x	x	x	x	x	x	x	x	x	√
Focus sectors	Agriculture and Forestry	√	√	√	√	√	√	√	√	√	√	√	√	√	√	√	√	√	√	√	√
	Food	x	√	√	√	√	√	√	√	√	√	√	√	√	√	√	√	√	√	x	x
	Energy	√	√	√	√	√	√	√	√	√	√	√	√	√	√	√	√	√	√	√	√
	Pulp and paper	x	√	√	√	√	√	x	x	√	√	x	√	x	√	x	√	√	√	√	x
	Wood-processing	x	√	√	√	√	√	x	x	√	√	√	√	√	√	√	√	√	√	x	x
	Chemical industry	√	√	√	√	√	√	√	√	√	√	√	√	√	x	√	√	√	x	√	√
	Medical industry	√	√	√	x	x	√	√	√	√	√	x	√	√	x	√	√	x	√	√	√
Targets	Vision and general target	x	√	x	√	√	√	x	√	x	√	x	x	x	√	x	√	x	(√)	x	√
	Measurable targets	x	x	x	√	√	√	√	√	√	√	√	√	√	√	√	√	√	√	√	√
Implementation focus	R&D	x	√	√	√	√	√	√	√	√	√	√	√	√	√	√	√	√	√	√	√
	Transition to market	x	x	√	√	√	√	√	√	x	√	√	x	√	x	√	√	√	√	√	√
	Policies	x	x	√	√	√	√	√	√	√	√	√	√	x	√	√	√	√	x	√	x

√, applicable; x, not applicable; (√), no governmental bioeconomy strategy, but high governmental attention (eg, national blueprint).

Note: No responses received from country representatives of South Korea and Ireland.

AU, Australia; AT, Austria; BE, Belgium; BR, Brazil; CA, Canada; HR, Croatia; DK, Denmark; FI, Finland; FR, France; DE, Germany; IT, Italy; JP, Japan; NL, Netherlands; NZ, New Zealand; NO, Norway; ZA, South Africa; SE, Sweden; CH, Switzerland; UK, United Kingdom; US, United States.

The focus of the survey was on official governmental bioeconomy policy positions, with regional and industry strategies serving as additional evidence for the current state of bioeconomy development in a country. Strategy documents provided by the IEA Bioenergy country representatives were analyzed in a framework of questions to compare patterns as the definition and scope of bioeconomy, vision and (measurable) targets, economic sectors in the focus of the strategies, current focus of implementation, and the position of bioenergy in a future bioeconomy.

It was found that the following five countries have official governmental bioeconomy policy strategies (with typical elements as objectives, focus of action, activities/measures, targets):

- Finland: The Finnish BioEconomy Strategy (2014)
- Germany: National policy strategy BioEconomy (2014), National research strategy BioEconomy 2030 (2010)
- The Netherlands: Framework memorandum on the Biobased Economy (2012)
- South Africa: The BioEconomy Strategy (2013)
- Sweden: Swedish Research and Innovation Strategy for a Biobased Economy (2012).

Furthermore, the following eight countries had governmental interest in the topic of bioeconomy, although no national strategy yet existed, but national blueprints, green economy strategies, or strategies for a biobased industry:

- Austria: Research, Technology and Innovation Strategy for Biobased Industries in Austria (2014)
- Denmark: Growth plan for water, bio and environmental solutions (2014)
- France: Non-food uses of biomass (2012)
- Japan: Biomass Industrialization Strategy (2012)
- Norway: Research Programme on Sustainable Innovation in Food and Biobased Industries "BIONÆR" (2012–2022) (2012)
- Switzerland: Green Economy: Report and Action Plan (2013)
- United Kingdom: Bioenergy Strategy (2012), Strategy for Agricultural Technologies (2013).
- United States: National Blueprint Bioeconomy (2012).

Additional findings of the study include that bioeconomy development is almost always a top-down, policy-driven process. A regional, bottom-up development approach was found in Australia, Belgium, Canada, Japan, and Sweden. Most strategies formulate a bioeconomy vision and general targets, and measurable policy targets have been defined in strategy documents of Canada, Finland, the Netherlands, and the US. Research and development (R&D) measures in all countries focus on sustainable biomass supply and bioenergy production. In most of the countries (80% of those polled) the chemical sector is identified as a priority area for the transition to a bioeconomy. The energy sector is important in all national transition strategies, in most (65% of those polled) cases biomass for bioenergy has an equal position to other economic sectors in the bioeconomy. Bioenergy as future priority use of biomass was found in Australia, Brazil, Denmark, Italy, Japan, and the US. Biorefining is often identified as a key technology for successful bioeconomy deployment. For market deployment, all countries focus on research and development, 70% on measures for transition to market, and 25% on additional policy development.

1.3 SCOPE, OBJECTIVE, AND OUTLINE

Despite the vast amount of politically driven strategies, there is still little understanding on how current markets will transition towards a national and essentially global bioeconomy. This joint analysis brings together expertise from three IEA Bioenergy subtasks, namely Task 34 on Pyrolysis, Task 40 on International Trade and Markets, and Task 42 on Biorefineries. The underlying hypothesis of the work is that bioeconomy market developments can benefit from lessons learned and developments observed in bioenergy markets. The question is not only how the bioeconomy can be developed, but also how it can be developed sustainably in terms of economic (eg, risk reduction and piggybacking on existing industry) and environmental concerns (eg, nonfood biomass-based). The strength of bringing three IEA Bioenergy subtasks into this analysis is found in each task's area of expertise. Tasks 34 and 42 identify the types of biorefineries that are expected to be implemented and the types of feedstock that may be used. Task 40 provides complementary work including a historical analysis of the developments of biopower and biofuel markets, integration opportunities into existing supply chains, and the conditions that would need to be created and enhanced to achieve a global biomass trade system supporting a global bioeconomy. It is expected that a future bioeconomy will rely on a series of tradable feedstock intermediates, that is, commodities. Investigating the prerequisites for such a commoditization, and lessons learned by other industries play a central role in this analysis.

The analysis covers an overview of biorefineries in a global biobased economy, identifies feedstock and conversion pathways, and outlines the status of demonstration plants and underlying economics. It brings together lessons learned and case studies from the biopower and biofuel markets and covers a brief historical description of international bioenergy trade and markets and links these and future developments to biomass preprocessing options. Furthermore, it bridges current to future bioeconomy-related markets by identifying and describing logistical integration opportunities. Several case studies of existing supply chains in the bioenergy markets are analyzed with respect to increased volume and end-use markets. Chapter "Commoditization of Biomass Markets" analyzes market factors necessary for commoditization. The analysis also portrays potential transition strategies from the current, conventional bioenergy markets, to advanced, large-scale, increased-volume trade required for a global bioeconomy. The emergence or deployment of the future bioeconomy will depend on the ability to achieve commodity-type, tradable feedstock intermediates. The transition towards such a system will need to bridge logistical as well as market structures. This analysis provides suggestions on how this could be done and what prerequisites are necessary.

The report starts with an overview of biorefineries in a global bioeconomy, identifying feedstock and conversion pathways (see chapter: Development of Second-Generation Biorefineries), and outlining the status of demonstration plants and underlying economics (see chapter: Biorefineries: Industry Status and Economics).

Chapter "Sustainability Considerations for the Future Bioeconomy" presents sustainability aspects that relate to biomass use within the future bioeconomy. It draws upon lesson learned from the biofuel and food systems.

Chapter "Biomass Supply and Trade Opportunities of Preprocessed Biomass for Power Generation" follows, bringing together lessons learned and case studies from the biopower and biofuel markets. It also covers a brief historical description of international bioenergy trade and markets and links these and future developments to biomass preprocessing options.

Chapter "Commodity Scale Biomass Trade and Integration With Other Supply Chains" bridges current to future bioeconomy-related markets by identifying and describing logistical integration opportunities. Several case studies of existing supply chains in the bioenergy markets are analyzed with respect to increased volume and end-use markets.

Chapter "Commoditization of Biomass Markets" analyzes market factors necessary for commoditization. It is based on case studies of the grain industry and financial markets.

The book closes with chapter "Transition Strategies—Resource Mobilization Through Merchandisable Feedstock Intermediates," which details potential transition strategies from current, conventional bioenergy markets, to advanced, large-scale, increased-volume trade required for a global bioeconomy.

REFERENCES

Beermann, M., Jungmeier, G., Pignatelli, V., Monni, M., Ree, R.V., 2014. National BioEconomy Strategies—IEA Bioenergy Implementing Agreement Countries. Joanneum Research, ENEA, Wageningen University, Graz, Austria, Available from: <http://www.iea-bioenergy.task42-biorefineries.com/upload_mm/9/d/b/91051282-061a-4d6b-8822-865428a42038_BioEconomy%20Survey%20IEA%20Bioenergy%20IA%20Countries_website.pdf>.

Bell, G., Schuck, S., Jungmeier, G., Wellisch, M., Felby, C., Jorgensen, H., et al. 2014. IEA Bioenergy Task 42 Biorefining: Sustainable and Synergetic Processing of Biomass Into Marketable Food & Feed Ingredients, Chemicals, Materials and Energy (Fuels, Power, Heat). Wageningen, IEA Task42: 63.

EC, 2012. Innovating for Sustainable Growth: A Bioeconomy for Europe. European Commission, Brussels, Belgium, Available from: <http://ec.europa.eu/research/bioeconomy/pdf/201202_innovating_sustainable_growth_en.pdf>.

IEA, 2014. World Energy Outlook. International Energy Agency, Paris.

IPCC, 2007. Climate Change 2007: Mitigation of Climate Change. Contribution of Working Group III to the Fourth Assessment Report of the Intergovernmental Panel on Climate Change. Cambridge University Press, Cambridge, UK and New York, USA.

IPCC, 2014. Climate Change 2014: Mitigation of Climate Change. Contribution of Working Group III to the Fifth Assessment Report of the Intergovernmental Panel on Climate Change. Cambridge University Press, Cambridge, United Kingdom and New York, NY, USA.

Langeveld, J.W.A., Dixon, J., Jaworski, J.F., 2010. Development perspectives of the biobased economy: a review. Crop. Sci. 50 (Suppl. 1), S-142–S-151.

NEA, 2014. Setting Up International Biobased Commodity Trade Chains. A Guide and 5 Examples in Ukraine. Netherlands Enterprise Agency, Den Hague, the Netherlands, Available from: <http://english.rvo.nl/sites/default/files/2014/06/Setting%20up%20international%20biobased%20commodity%20trade%20chains%20-%20May%202014.pdf> (accessed 04.08.14.).

OECD, 2009. The Bioeconomy to 2030: Designing a Policy Agenda. Organisation for Economic Co-operation and Development, Paris, France, 322, Available from: <http://www.oecd.org/futures/long-termtechnologicalsocietalchallenges/thebioeconomy-to2030designingapolicyagenda.htm>.

Trainer, T., 2010. Can renewables etc. solve the greenhouse problem? The negative case. Energy Policy 38 (8), 4107–4114.

CHAPTER 2

Development of Second-Generation Biorefineries

H. Stichnothe[1], H. Storz[1], D. Meier[2], I. de Bari[3] and S. Thomas[4]

[1]Thünen Institute of Agricultural Technology, Braunschweig, Germany
[2]Thünen Institute of Wood Research, Hamburg, Germany
[3]Division of Bioenergy, Biorefinery and Green Chemistry, ENEA Centro Ricerche Trisaia, Policoro, Italy
[4]US Department of Energy, Bioenergy Technologies Office, Golden, CO, United States

Contents

Abstract

A wide range of nonfood biomass and conversion technologies can be used for the production of bioenergy and biobased products. The fermentation of lignocellulosic-derived sugars and the thermochemical conversion of biomass (eg, fast pyrolysis) are examples of relevant conversion technologies. The main product of fast pyrolysis is bio-oil, which can be used directly in stationary boilers or after upgrading as a drop-in blend component in existing refineries. Bio-oil requires chemical upgrading, before it is suitable as fuel. The commercial use of bio-oil for material/chemical purposes is currently limited to minor food uses (ie, smoke aroma and flavor enhancers). Different pretreatment technologies can be used in the initial conversion of biomass to sugars for fermentation. Technical obstacles in those pretreatment processes differ among the various approaches, but can include insufficient separation of cellulose and lignin, formation of byproducts that inhibit downstream fermentation, high use of chemicals and/or energy, as well as high costs for cellulase enzymes, although the latter has decreased substantially in recent years. There is currently no consensus on a preferred pretreatment method or combination of methods. A wide range of biofuels and biobased chemicals can be produced from sugars via fermentation and/or chemical

© 2016 Elsevier Inc.
All rights reserved.

conversion, including advanced biofuels and chemical intermediates. The integration of different pretreatment and conversion technologies in biorefineries can maximize the use of biomass components and improve the efficiency of the entire value chain. In the mid- to long term, thermochemical and biochemical conversion of lignocellulosic biomass are promising technologies for the production of biofuels and biobased chemicals.

2.1 INTRODUCTION

The use of conventional crops for biofuels and bioenergy is controversial due to direct and indirect land use change perceptions. There are additional concerns, including increases in crop and food prices in developing countries caused by competition for land (Schmidhuber, 2007; Fargione et al., 2008; Searchinger et al., 2008; Bringezu, 2009; Kim et al., 2009, 2012; Dale and Kim, 2011; Kim and Dale, 2011; Kline et al., 2011; O'Hare et al., 2011; Bringezu et al., 2012; Broch et al., 2013). Therefore, the use of nonfood biomass for the production of bioenergy and biobased chemicals is preferred.

A wide range of nonfood biomass and conversion technologies can be used for the production of bioenergy and biobased products. The availability of nonfood biomass is dependent on a number of regional factors. The amount of organic waste and processing residues (eg, from the food industry) depends on the livestock density, and the type and scale of the food industry. In Europe the organic waste and residue availability for bioenergy production is estimated to be in the range of 550 million tonnes[1] per year (t/year), of which 375 million t (~68%) are agricultural residues (Searle and Malins, 2014). The amount of surplus agricultural residues in India is approximately 235 million t/year (Hiloidhari et al., 2014) and in China 500 million t/year (Jiang et al., 2012). In the United States (US), primary and secondary agricultural residues are conservatively estimated to be 240 million dry t/year in 2030 at a farm gate price of $66 per metric t or 114 million dry t/year at a price of $40 per metric t (US-DOE et al., 2011). However, the amount of potentially harvestable residues must always take into account the need to avoid any depletion of soil organic carbon stocks (Montanarella and Vargas, 2012). Primary and secondary forest residues from non-Federal land in the US are projected to be at 86 million dry t/year in 2030 at a forest landing price of

[1] One metric tonne = 1000 kg = 1 Megagram (Mg).

$66 per dry t or 72 million dry t/year at a price of $40 per t (US-DOE et al., 2011). Dedicated woody and herbaceous energy crops are also useful resources for biorefineries and approximately 342 million dry t/year are projected to be available in the US in 2030 at a farm gate price of $66 per t or 30 million dry t/year at a price of $40 per t (US-DOE et al., 2011). The high cost of establishment of these perennial crops and the lack of a harvestable crop in the first year or two requires a higher farm gate price in order for farmers to be able to make money growing these crops.

The focus of this chapter is mainly on technologies that enable the conversion of nonfood biomass into value-added products. We attempt to identify the most promising technology options and/or combinations for biomass processing.

2.2 TECHNOLOGY AND FEEDSTOCK MATRIX

Fig. 2.1[2] identifies technologies that facilitate the conversion of a variety of biomass materials to value-added products. However, the main focus here is on the conversion of woody and herbaceous lignocellulosic (nonfood) materials, including agricultural and forestry primary and secondary residues, dedicated energy crops, and the organic fraction of municipal solid waste. Oils and fats are important feedstock of the chemical industry in the past and in the present. Many newly produced biobased chemicals are from sugars and due to the food–fuel debate lignocellulosic-derived sugars (also called second-generation sugars) have recently gained much attention. Thus, this chapter will be focused on the process and technologies for the production of cheap and clean second-generation sugars.

As Fig. 2.1 indicates, there are many process technology options by which a variety of edible and nonedible biomass materials can be converted into useful bioenergy, fuels, and chemicals. The production of heat and power (eg, combustion in district heating systems, waste incineration, or large-scale cocombustion) are mature technologies and thus not discussed here. Anaerobic digestion (AD) and composting (indicated in gray boxes in Fig. 2.1) are briefly described as they are relevant to agricultural residue treatment, the recycling of nutrients, and thus sustainable land management. The main focus of this chapter is the conversion of biomass into value-added products via new or emerging technologies that are indicated by black boxes in Fig. 2.1.

---

[2]Technologies discussed in detail are emphasized in bold.

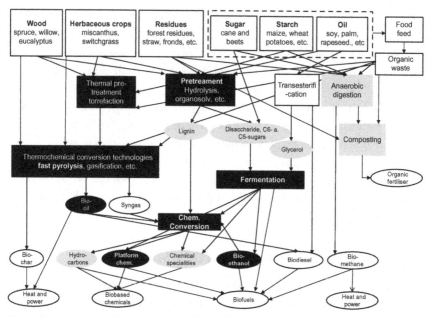

Figure 2.1 Overview of conversion technologies for biomass.

2.2.1 Composting and Anaerobic Digestion

Composting is a biological process in which organic waste such as food waste, manure, garden waste, trimmings, etc., are converted into a humus-like substance by microorganisms under aerobic conditions. Composting is a robust, low-cost, and low-tech option that can handle a variety of organic wastes varying in composition and moisture content. There are three major designs used in composting: static piles, vessels, and windrows. Windrow composting is usually considered the most cost-effective composting option. The organic material is layered into long piles and thoroughly mixed by turning machines. The temperature within windrows can reach 65°C; hence composting can also act as a biological drying process. Therefore, composting is frequently used in combination with biogas production in areas with high livestock density, where surplus nutrients cause considerable problems (Luo et al., 2013). Compost is used as a soil amendment and slow-release organic fertilizer. During the composting process, a large part of the carbon in the organic waste feedstock is metabolized to carbon dioxide, which is usually not captured for recycle into a useful bioproduct, but it provides a low-tech method to reduce the volume of waste streams and provide a useful product.

Compost is one of the best sources of stable organic matter from which new humus can be formed in degraded soils. An estimated 45% of European soils have low organic matter content, principally in southern Europe, but also in areas of France, the UK, and Germany. Standards on the use and quality of compost exist in most countries (eg, Organic Farming Regulation or eco-labels for soil improvers). The standards differ substantially, partly due to differences in national and international soil policies.

In AD processes, organic material is converted in the absence of oxygen into raw biogas (methane plus carbon dioxide) and unmetabolized solids, or digestate, which can be used as fertilizer. After purification, the methane component of biogas (ie, biomethane) is equivalent in energy content to fossil natural gas, and can be used as fuel or as a source of biobased hydrogen or syngas. The carbon dioxide component rejected during the biomethane purification process is generally exhausted to the atmosphere and not captured for reuse. Codigestion facilities are typically agricultural anaerobic digesters that simultaneously accept two or more input materials. Feedstock for AD can include a range of organic materials, including biodegradable waste, such as grass clippings, food waste, sewage, and manure. Anaerobic digesters can also be fed with dedicated energy crops, such as *Silphium perfoliatum* or silage maize, for biogas production, although this is generally considered to be controversial due to land use issues. Lignocellulosic biomass is not the preferred feedstock for AD, because most anaerobes are unable to degrade lignin. However, the carbohydrate content of plant cell walls is generally useable by AD microbial consortia.

2.2.2 Preprocessing Technologies

Important issues in the handling and delivery of biomass include low bulk and energy densities and instability in storage, that necessitate conditioning or preprocessing operations prior to introduction into a conversion process. In addition, biomass delivered to the conversion process must possess physical and chemical properties that fall within the design specifications of the biomass conversion process, which vary from one conversion process to another. Preprocessing technologies of potential interest include— but are not necessarily limited to—operations such as drying, milling, fractionation and separation, blending, densification, and torrefaction.

2.2.2.1 Basic Biomass Preprocessing Methods

Chemical-physical properties of interest include moisture, ash, carbohydrates, lignin, and energy content, among others. Moisture in biomass presents problems from several points of view. Firstly, moisture content above 30% (by weight) permits microbial growth in aerobic and anaerobic conditions, which can result in significant mass losses ("shrinkage") and quality degradation over time in storage. Secondly, moisture in biomass contributes to the weight of material that must be transported from the field to the storage facility or biorefinery, which incurs undesirable transport costs. In addition, most thermochemical and combustion conversion processes prefer very dry (~10% moisture) feedstock materials. The moisture content of herbaceous materials is managed with established agronomic practices, which include chopping and drying crop residues, such as corn stover and wheat straw, in the field using solar radiation, or through natural senescence processes in perennial species, such as switchgrass and miscanthus.

Tropical species, such as energycane, and annual species, such as sweet and biomass sorghum types, that do not naturally senesce and dry in the field may require a just-in-time harvest and delivery strategy, such as that which the sugarcane industry employs today around the world. Woody species, such as hybrid poplar, shrub willow, eucalyptus, and loblolly pine are typically harvested green and usually have a moisture content around 50% at harvest. Partial drying of whole trees in the field after felling is a possibility for some woody species, depending on the harvest strategy. Mechanical drying may be necessary in some cases to preserve biomass quantity and quality in storage, and also to prevent catastrophic spontaneous combustion events due to heating due to microbial activity, but it is extremely energy-intensive, and therefore expensive.

Ash content can also present problems for the biomass conversion industry. In general, ash content should be as low as possible for any conversion process, and therefore harvest, collection, and storage methods should take care not to introduce additional ash from soil contamination. Debarked wood can be very low in ash content (eg, <1% by dry weight). Bark contains significant ash and is also high in lignin, so significantly increases the ash content of whole stem wood. Harvested trees that are skidded to a landing prior to chipping or loading onto trucks can acquire additional ash through contamination by soil. By contrast, very clean herbaceous materials typically contain 4–7% ash, which is significantly higher than clean wood chips, and makes them more suitable for biochemical conversion processes.

As a result of their low ash content and higher energy content, clean, dry wood chips are generally preferred for thermochemical conversion processes over any herbaceous feedstock. Mineral ash contributes nothing to any biomass conversion process yield other than a potential fertilizer coproduct, and generally tends to interfere with conversion processes by reducing yield per tonne of input biomass. The ash content of herbaceous materials can also significantly increase abrasive wear in the moving parts of processing machinery. In thermochemical processes ash can cause slagging in thermochemical reactors and fouling of downstream chemical catalysts. Some mineral species are worse than others in thermochemical processes. In extreme situations, some sort of mechanical or chemical preprocessing step (ie, sieving; washing) may be warranted to reduce the ash content of feedstock materials.

High carbohydrate content is especially important in biochemical conversion processes, as all biofuels and bioproducts are derived from the sugars in cell wall carbohydrates. Due to the oxygen-rich composition of carbohydrates, they have a lower energy content per unit mass than lignin, and are less desirable feedstock components for thermochemical conversion processes.

High lignin content is especially important for thermochemical conversion processes. Lignin has a higher energy content per unit mass than carbohydrate because it is largely comprised of energy-rich aromatic rings and has a lower oxygen content than carbohydrate.

Examples of relevant physical properties include biomass particle size and shape, particle size distribution, fibrous nature, and density. A nominal particle dimension is a typical specification for feedstocks entering a biomass conversion process. A variety of milling processes can be used to achieve this specification, including hammer milling, knife milling, ball milling, each of which has its advantages and disadvantages, depending on downstream needs. A narrow size distribution range is generally more desirable than a broad size distribution profile around the specified nominal dimension. However, a narrow size distribution can be difficult to achieve, given the mechanical properties (brittleness versus elasticity) of biomass materials, and the relative crudeness of most industrial milling processes. In particular, very fine and very coarse particle size fractions are to be avoided, but for different reasons. Larger particles can occlude openings in solid and slurry storage and conveyance systems, while very fine particles tend to settle as sludge once exposed to aqueous conditions. Both of these outcomes can cause process interruptions and considerable downtime. Sieves can be used to fractionate milled materials to remove fines, as well as oversized particles, if necessary.

Recent investigations demonstrate that particle sizing can impact the economics of the production of second-generation sugars via pretreatment and enzymatic hydrolysis. Among the available pretreatment technologies, thermochemical pretreatments have exhibited a wide range of particle sizes below which no increase in pretreatment effectiveness was observed from <0.15 to 50 (Vidal et al., 2011). However, the optimal particle size varies for different combinations of pretreatment technologies and feedstock.

2.2.2.2 Densification and Thermal Pretreatment

Highly fibrous materials tend to be difficult to transport and convey, due to their tendency to tangle, resulting in bridging and blockage of orifices and passageways. Densification of biomass is an option to overcome those obstacles. Densification processes include pelletization and briquetting, which transform loose, fluffy biomass into reasonably uniform (in size, shape, and density), smooth materials that flow evenly and well in standard storage and conveyance equipment.

The combined use of torrefaction and pelletization presents an opportunity to improve stability of biomass in storage, reduce transportation costs, and improve the efficiency of handling operations. During torrefaction, biomass is heated to approximately 200°C in order to reduce and/or remove its water content and drive off volatile compounds. Torrefaction typically reduces the energy content of the torrefied material to approximately 90% of the original material (ie, about 10% of the original biomass is lost during torrefaction). The resulting torrefied material is porous, which provides an opportunity for undesirable moisture uptake, so densification via pelletization is an attractive option.

2.2.3 Pretreatment: Physical, Chemical, and Biochemical

In connection with biocatalytic conversion processes, the term pretreatment is used for a number of single or combined physical, chemical, and biological processes that are used to efficiently convert lignocellulosic biomass, which is usually recalcitrant to enzymatic hydrolysis, into an activated form that can be cost= effectively hydrolyzed enzymatically to produce C5 and C6 sugars in aqueous solution. Thus, the ultimate pretreatment technology would be a low-cost, sustainable process that renders lignocellulosic biomass completely accessible to enzymes and/or microorganisms that efficiently and completely hydrolyze all carbohydrate polymers present in plant cell walls and avoids the formation of toxic degradation products that negatively impact downstream biological or chemical processing, and thereby process yield and profitability.

Pretreatment technologies for biocatalytic conversion technologies can be broadly divided into two categories: pretreatment and fractionation.[3] The goal of fractionation is to separate biomass into its three main components: hemicellulose, cellulose, and lignin. The quality of the manufactured cellulose can be of similar quality as obtained by conventional pulping processes. Although many research groups have tried to find a more economically attractive application for lignin, it is usually used as fuel for the conversion of cellulose- and hemicellulose-derived sugars into value-added products, which can be energy-intensive. Hemicellulose is commonly treated as a secondary sugar stream from lignocellulosics-based biorefineries. This is often due to the lack of efficient and robust fermentation processes for the conversion of C5 sugars to bioethanol or other product. The quality of the hemicellulose stream, as defined by its sugar composition and the level of sugar degradation byproducts, depends on the specific biomass pretreatment employed. In this regard, optimized pretreatment and fractionation schemes are necessary to ensure the maximum exploitation of the biomass carbohydrates.

The development of pretreatment technologies for lignocellulose is driven by the production of second-generation bioethanol. Therefore pretreatment methods are the subject of intensive research activities (Mosier et al., 2005; Yang and Wyman, 2008; Gírio et al., 2010; Goh et al., 2011; Shi et al., 2011; Panagiotopoulos et al., 2012; Chiesa and Gnansounou, 2014).

One example of a pretreatment strategy is dilute acid hydrolysis, which is probably the oldest pretreatment option available. Dilute acid pretreatment processes typically employ a dilute (eg, 0.1%) strong acid, such as sulfuric acid, high temperature (eg, 150–200°C), and a fairly short residence time (eg, 10 minutes) to break the lignin seal and nearly completely hydrolyze hemicellulose into its constituent sugars. The net effect is to increase the surface area of cellulose fibers that is accessible to subsequently added cellulase enzymes. This approach ultimately results in increasing the concentration and amount of fermentable sugars from a given amount of input biomass.

There are a variety of physical and physio-chemical methods that have advantages and disadvantages (Linde et al., 2008; Hendriks and

---

[3] Fractionation of biomass is conducted by the PureVision process or Organosolv methods (Organosolv, NREL Clean Fractionation, etc.).

Zeeman, 2009; Alvira et al., 2010; Goh et al., 2011; Lai et al., 2014; Reisinger et al., 2014; Papa et al., 2015). In general, ammonia fiber explosion, wet oxidation, and liquid hot water are more suitable for agricultural residues and dedicated herbaceous energy crops, while steam explosion is applicable for both herbaceous and woody biomass. Hardwoods are typically less recalcitrant to dilute acid pretreatment than softwoods because their hemicelluloses are composed of acetylated xylans that rapidly hydrolyze in water at elevated temperatures. Another reason for the recalcitrance of softwood species is the presence of a more heavily crosslinked lignin in softwood cell walls relative to hardwoods and herbaceous species. Hardwoods also contain lower contents of readily fermentable C6 sugars in their hemicellulose than softwoods, as do agricultural residues and herbaceous crops (Gnansounou, 2008). This is because softwood hemicelluloses contain higher amounts of the C6 sugars, mannose and galactose, while hardwood and herbaceous species hemicellulose contains primarily C5 sugars, mainly xylose and arabinose. Furthermore, in acid pretreatment conditions pentose sugars are more easily degraded to byproducts, such as short-chain organic acids and furan compounds. Thus, the preferred pretreatment should facilitate high hemicellulose sugar recovery yields, while maintaining an acceptable downstream cellulose enzymatic hydrolysis rate and glucose yield.

Pretreatment processes are frequently classified as biological, physical, chemical, physicochemical, and solvent-assisted. Most chemical/biochemical pretreatments result in sugars in aqueous solution and a solid fraction. However, sugar degradation products are often also produced during pretreatment, which can negatively affect subsequent purification and/or conversion steps. The types and relative amounts of the sugars produced depend on process conditions and feedstock composition. Table 2.1 shows selected process conditions for various pretreatment methods as well as advantages and disadvantages of the methods.

2.2.4 Saccharification of Cellulose and Hemicellulose

The use of hydrolytic enzymes to degrade cellulose to fermentable sugar is generally considered more effective than the use of concentrated mineral acids, because enzymes are highly specific and can work at mild process conditions, for example, 45–50 °C and pH 4.8–5 for cellulases derived from the filamentous fungus, *Trichoderma reesei*. In addition, the mild chemical conditions employed for enzymatic processes obviate the need for extremely expensive, corrosion-resistant materials of construction

Table 2.1 Description of pretreatment methods for lignocellulosic biomass based on (Mergner et al., 2013)

		Main operating conditions	Advantages/disadvantages
Biological		Brown-, white- and soft-rot fungi at room temperature for several days	Low energy consumption/long incubation time (weeks). Low hydrolysis rates of polysaccharides are obtained after a few days of treatment
Physical	Chipping, grinding, milling	Room temperature and energy input $\leq 50\,kWh/t$	Increases surface area, reduces cellulose crystallinity/high energy consumption
Chemical	Dilute acid pretreatment	Acid loading in the range 0.1–2.5% w/w, T = 130–210°C	Solubilizes and depolymerizes hemicellulose/need to neutralize the product streams for downstream process steps/formation of sugar degradation products and fermentation inhibitors reduces product yield
	Alkaline hydrolysis	From 25°C to 170°C and treatment time from several hours/days to few seconds. Base-to-biomass: 0.075–0.5 g/g	Lignin and hemicellulose solubilization/need to wash the product/need to recycle base catalyst
	Ozonolysis	Gas flow rate 60 L/h, (ozone concentration 2.7% w/w), room temperature, 100–120 min	Reduces the lignin content/high cost of ozone
	Wet oxidation	170–200°C and treatment time from 60 to 3 min, O_2 5–12 bar, Na_2CO_3 3.4–3.6% (w/w)	Efficient removal of lignin/cost of chemicals

(Continued)

Table 2.1 (Continued)

		Main operating conditions	Advantages/disadvantages
Physio–chemical	Ammonia fiber explosion method	90–140°C for 5–45 min; NH_3/biomass DM (w/w) 1–1	High biomass destruction degree of cell wall architecture disruption without formation of a liquid fraction, or hydrolysis of any cell wall polymer/large amount of NH_3 needed/ammonia recycle required
	Steam explosion	160–260°C (0.69–4.83 MPa) for 5–15 min	High degree of cell wall architecture disruption, high solid content/partial degradation of C5 sugars
	Liquid hot water	120–220°C, 350–400 psi, for 10–20 min	High degree of hemicellulose solubilization/high water demand and dilution of sugar stream
	CO_2 explosion	150–185°C for 30–60 min; CO_2 120–200 psi	Does not require downstream processing/lignin is not affected by the treatment and high cost of supercritical fluids
Solvent fractionation	Organosolv	170–200°C for 30–60 min, alcohol (30–78%) with or without addition of catalysts (ie, NaOH, H_2SO_4)	Extraction of hemicellulose and low molecular-weight lignin/cost of solvent recycle
	Ionic liquid (ILs)	80–190°C for 24–30 min	High fractionation of lignin and polysaccharides/high cost of ILs requires high recycle efficiency

that are required when using concentrated mineral acids. Concentrated mineral acids also degrade sugars to unfermentable products, which results in yield losses.

Saccharification of lignocellulosic material requires the use of several enzymes with complementary activities: endoglucanase, which attacks regions in the interior of linear cellulose chains; exoglucanases or cellobiohydrolases, which hydrolyze cellobiose units from the ends of cellulose chains; and β-glucosidase, which converts cello-oligosaccharides and cellobiose into glucose. There are also several complementary enzymes that attack hemicelluloses, including glucuronidase, xylan acetylesterase, xylanase, β-xylosidase, galactomannase, and glucomannase. Filamentous fungi, such as *Trichoderma viride, T. reesei*, and *Trichoderma longibrachiatum*, are the principal producers of commercially available cellulases. These commercial products are cocktails containing mixtures of several enzyme types. Prior to their use in the production of second-generation biofuels, cellulases were included in detergent formulations and employed in the manufacture of "stonewashed" jeans.

The economics of biomass saccharification is sensitive to the cost of cellulase enzymes. The low specific activity of cellulases requires large amounts of enzyme to effectively saccharify the cellulose in biomass in a reasonable amount of time. In addition, unproductive binding of the enzymes to biomass increases the amount of enzyme required and therefore operational costs in the process. Addition of surfactants can be useful to decrease nonproductive binding of cellulases to lignin and polysaccharide–lignin complexes (Cui et al., 2011).

Enzyme companies and others have made significant progress toward reducing enzyme cost by streamlining enzyme production and formulation processes, inclusion of new enzymes in cellulase enzyme cocktails, and increasing the enzyme specific activity using enzyme engineering strategies. For example, with funding from the U.S. Department of Energy (US DOE), Genencor, Novozymes, and DSM each have dramatically reduced the cost and improved the quality of cellulase enzyme cocktails for use in cellulosic biomass conversion processes. Genencor has launched their Accellerase Duet product produced by a genetically modified strain of *T. reesei*, and which contains both cellulases and xylanases. Similarly, Novozymes has released their CELLIC product family (CTEC2 and CTEC3, and HTEC). These new cellulase products have been designed especially for use in the production of second-generation bioethanol. Dyadic is another US company that is producing cellulase enzyme

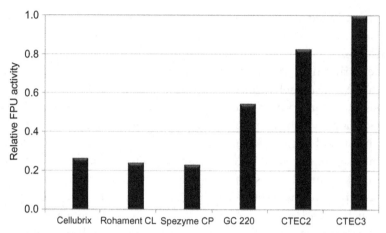

Figure 2.2 Activities of some commercial mixtures of enzymes normalized to the activity of the latest CTEC3.

cocktails for industrial use. Fig. 2.2 describes the improvement of the commercial blends.

Feedstock composition and the specific requirements of downstream processes determine the preferred pretreatment strategy. The types and relative amounts of sugars present in the feedstock are a function of the plant species, but the availability of those sugars for downstream conversion processes are a function of how well the pretreatment and enzymatic saccharification processes perform. Hence, the preferred combination of pretreatment and conversion technology depends on a number of factors and therefore no one-size-fits-all technology option currently exists. The ideal pretreatment process should result in high yields of fermentable sugars from hemicellulose and cellulose, avoid the generation of toxic compounds and yield losses due to sugar degradation, minimize enzyme usage and associated cost, and require little or no energy or chemical inputs. The following section describes biochemical conversion technologies, as well as their connection to pretreatment processes.

2.2.5 (Bio)-Catalytic Production of Bioethanol and Various Chemicals

Bioethanol has become the most commonly used biofuel worldwide. Bioethanol can be blended with gasoline or used in the synthesis of ethyl tertiary-butyl ether, which is used as an antiknocking agent in gasoline. Both compounds serve as oxygenates in gasoline, and were first used in the US to improve air quality in densely populated urban areas.

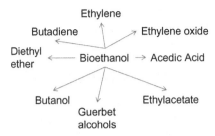

Figure 2.3 Chemicals and advanced fuels that can be produced from bioethanol using a variety of chemical methods.

Besides its use as a transportation fuel blending agent, ethanol is also considered a potent platform chemical for the synthesis of other value-added chemicals. Catalytic conversion of bioethanol has been successfully applied to the production of important chemicals such as ethylene, *n*-butanol, diethyl ether, etc. (Posada et al., 2013). Important bulk chemicals that can be derived from ethanol are shown in Fig. 2.3. The conversion of ethanol to bulk chemicals is still in its infancy, but this family of chemical conversions is probably one of the most promising near-term pathways to a biobased economy. Moreover, the conversion of ethanol to other chemicals would allow for the maintenance of existing downstream processing infrastructure. Ethanol production can be used as a proxy for other products derived from cellulosic sugars via biochemical conversion routes (eg, butanol, furan derivatives, etc.) because the equipment for purification and upgrading of those products is similar to that used for ethanol.

For the production of second-generation biofuels and biobased chemicals, hydrolysis and fermentation steps can be combined in simultaneous saccharification and fermentation (SSF) of hexoses or simultaneous saccharification and cofermentation (SSCF) of both hexoses and pentoses. Consolidated bioprocessing (CBP) is a system in which cellulose enzyme production, substrate hydrolysis, and fermentation take place simultaneously in the same reactor vessel using a fermentation organism that produces the enzymes that saccharify the cellulose. CBP offers the potential for lower biofuel production costs due the process integration, which eliminates the capital and operating costs associated with the separate production of enzymes and ensures higher efficiencies than separate hydrolysis and fermentation (SHF) or SSF (van Zyl et al., 2013). The *Saccharomyces cerevisiae* strain developed by the Mascoma Corporation is the best CBP organism engineered so far, as this strain can convert several cellulosic substrates to ethanol with addition of minimal exogenous enzymes (Mcbride,

Table 2.2 Top chemicals derived from biomass identified by the US Department of Energy (US DOE)

List in 2004 (Werpy et al., 2004)	List in 2010 (Bozell and Petersen, 2010)
3-Hydroxybutyrolactone	Biohydrocarbons
Aspartic acid	Ethanol (C2)
Glycerol	Lactic acid (C3)
Glucaric acid	Furans
Glutamic acid	
3-Hydroxypropanoic acid	
Itaconic acid	
Levulinic acid	
2,5-Furan dicarboxylic acid	
Sorbitol	
C4 acids (succinic, fumaric and malic acid)	
Xylitol (C5)	

2010). Various recombinant microorganisms have demonstrated impressive saccharification potential (Ha et al., 2011), although their use for large-scale industrial processes still requires fine-tuning of the entire process. The (simultaneous) fermentation of mixed C5 and C6 sugar hydrolysates is still one of the most important challenges in the production of biofuels and biobased chemicals.

Apart from bioethanol, a number of platform chemicals and advanced biofuels can be produced from sugars derived from lignocellulosic materials. The US DOE identified the top 12 chemicals potentially derived from biomass in 2004 (Werpy et al., 2004). The list was extended in 2010 to include alcohols, acids, amino acids, and terpenes (Bozell and Petersen, 2010). Table 2.2 lists chemicals that can be derived from sugars either through fermentation, for example, C4 acids or chemical conversion, for example, furan derivatives produced via catalyzed dehydration of sugars.

Succinic acid, a C4-dicarboxylic acid platform chemical, has received increasing attention for the production of new polyesters with good mechanical properties combined with full biodegradability. Succinic acid is widely used, especially for the production of renewable polyesters. Succinate concentration as high as 110 g/L has been achieved from glucose by the rumen organism *Actinobacillus succinogenes* (Liu, 2000). It can also be produced by *Anaerobiospirillum succiniciproducens* using glucose, lactose, sucrose, maltose, or fructose as carbon sources. The reduction of the carboxylate groups is one of the available routes to 1,4-butandiol (BDO), a building block for the production of various polyesters.

Several other organic acids are also considered to be potential platform molecules. *Lactic acid* is probably the most prominent. It can dimerize and subsequently form lactide, an intermediate for polymerization to the biodegradable plastic, polylactic acid. The monomer acid is typically produced by glucose fermentation using *Lactobacillus delbrueckii* (Datta, 2006). Luo et al. (1997) indicated that using the SSF strategy for lactic acid production offers the advantage that enzymatic hydrolysis and lactic acid fermentation work at similar process conditions (eg, temperature and pH). Other organic acids may be a suitable platform for specialty chemicals or niche markets, but the large volumes of waste products produced (ie, gypsum) are problematic and require the development of a more efficient process design.

Other biobased chemicals and advanced fuels obtained through the fermentation of sugars include alcohols higher than ethanol, diols, microbial lipids, and advanced hydrocarbons. Some of these are described below.

Biobutanol has a higher energy density and is less hydrophilic than ethanol, both of which make it a better blending agent for gasoline than ethanol. It can be produced by acetone-butanol-ethanol (ABE)–type anaerobic fermentation organisms (Atsumi et al., 2008). *Clostridium beijerinckii* ATCC 55025 can utilize hexoses and pentoses simultaneously to produce butanol (Liu et al., 2010). The main problem associated with the industrial production of biobutanol is the high energy required for purifying the product from fermentation liquors at low butanol concentrations. The combination of fermentation with pervaporation has been proposed to separate butanol during fermentation and increase the process efficiency (Qureshi et al., 2001). Gevo and Butamax have successfully converted cellulose-derived isobutanol produced via a fermentation process into isobutylene and paraffinic kerosene (jet fuel) via well-understood chemical processes.

2,3-Butandiol (BDO) is a promising compound with three stereoisomer forms. It can be used both as liquid fuel and as a platform chemical. Both the levo- and dextro-isomers of BDO are chiral components for asymmetric synthesis reactions. Various microorganisms, namely *Klebsiella pneumoniae* (Ma et al., 2009), *Klebsiella oxytoca* (Cheng et al., 2010), *Paenibacillus polymyxa* (Gao et al., 2010), and *Enterobacter aerogenes* (Zeng, 1991), have been tested for the production of BDO by using various glucose sources. BDO can be converted to 1,3-butadiene, an intermediate for the synthesis of rubbers, polyesters, and polyurethanes, or to methyl-ethyl-ketone, a solvent and commonly used fuel additive.

Among polyalcohols, *xylitol*, a natural noncaloric sweetener with anticariogenic properties, has been considered as the main alternative to ethanol produced from C5 sugars. It is currently produced from the nickel-assisted hydrogenation of xylose. Sorbitol can also be biotechnologically produced by the bacterium *Zymomonas mobilis*, which uses fructose and glucose (Silveira, 2002).

Microbial lipids (predominantly triglycerides) obtained through the fermentation of lignocellulosic hydrolyzates with oleaginous microorganisms can be used for the production of free fatty acids, fatty alcohols (Metzger, 2006), and fatty acid methyl (or ethyl) esters, which can be further converted to several intermediates for the production of biodiesel, hydroprocessed esters and fatty acids, lubricants (Chowdhury, 2013), and surfactants. However, the commercialization of microbial lipids is still cost-prohibitive. More efficient and integrated processes are necessary for better techno-economics of this process (Gong et al., 2013).

Advanced biohydrocarbons are similar to conventional hydrocarbon fuels such as gasoline, diesel or jet fuels, but are produced from biomass feedstocks. Recently, Amyris engineered the isoprene metabolic pathways in some microorganisms for the production of terpenoids such as farnesene, a 15-carbon branched hydrocarbon that can be chemically hydrogenated to a drop-in diesel or jet fuel component (farnesane).

The major part of the current chemicals-based biorefineries use first-generation sugars but very little second-generation sugars. In Table 2.3, chemicals derived from lignocellulosic material are summarized together with selected process conditions.

2.2.5.1 Chemicals From Glycerol

The production of first-generation biodiesel will continue, particularly in Malaysia and Indonesia. Combining palm oil and biodiesel production would produce refined palm oil, biodiesel, and surplus electricity as primary products, as well as bleaching earth, palm fatty acid distillate (PFAD), and glycerol as byproducts. PFAD is a light brown solid at room temperature, which currently is used for soap and animal feed, but also could serve as feedstock for the oleochemical industry. PFAD also contains a significant amount of extractable vitamin E (Santosa, 2008).

Purified glycerol from biodiesel production is already a marketable product. However, vastly increased biodiesel production could lead to significant glycerol quantities that the market might not be able to absorb without causing a drop in glycerol price. If the most recent biodiesel

Table 2.3 Process conditions for chemicals derived from lignocellulosic material

Value-added product	Feedstock	Fractionation technology	Polysaccharide degradation	Micro-organism	Yield % or concentration	References	Applications
Ethanol	Sugarcane bagasse	Steam explosion	Enzymatic hydrolysis	*S. cerevisiae*	56.3 g/L	Amores et al. (2013)	Fuel, chemicals
Butanol	Wheat straw	Dilute sulfuric acid hydrolysis	Enzymatic hydrolysis	*Clostridium beijerinckii P*	0.44 g/g sugars; 0.36 g/L per h	Qureshi et al. (2008)	Fuel, chemicals
Xylitol	Wheat straw		Acid hydrolysis	*Candida tropicalis* (AS2. 1776)	42% (xylitol/xylose)	Zhuang et al. (2009)	Sweetener
Lactic acid	Water hyacinth	Na_2SO_3 Pretreatment in NaOH	Enzymatic hydrolysis	*Lactobacillus acidophilus*	0.86 g Lactic acid/g sugars consumed	Idreesa et al. (2013)	Personal care, bioplastics
Biosuccinic acid	Cotton stalks	Steam explosion	Enzymatic hydrolysis	*Actinobacillus succinogenes*	64% total sugars	Li et al. (2013)	Surfactant, foaming agent, antimicrobial agent, etc.
Triacyl-glycerides	Corn stover	Mechanical pretreatment	Enzymatic hydrolysis	*Cryptococcus curvatus*	86 mg lipid/g feedstock; 7.4 g/L	Gong et al. (2013)	Fuel
Glutamic acid	Corncob fibers hydrolysates		Enzymatic hydrolysis	*Bacillus subtilis HB-1*	24.92 g/L	Zhu et al. (2013)	Food, medical, cosmetics industries, etc.

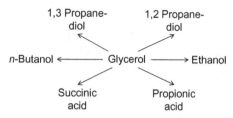

Figure 2.4 Promising chemicals from glycerol.

targets are fulfilled in Indonesia, then the amount of glycerol will increase from 10,000 t in 2009 to 750,000 t in 2016. Vlysidis et al. (2011) conducted a techno-economic analysis of biodiesel refineries. Their results indicate the importance of glycerol as a key building block for the production of commodity chemicals and as feedstock for various fermentation processes (Fig. 2.4).

Another option for the utilization of glycerol is the catalytic dehydration of glycerol to acrolein, a versatile intermediate. Acrolein derivatives, such as acrylic acid, an ingredient for paints and coatings, offer high industrial value. Currently, acrolein is mostly used for the synthesis of methionine. Methionine is a sulfur-containing amino acid that is required as a supplement in animal feed. Methionine has a global annual demand of approximately 850,000 t (Willke, 2014) and an expected annual growth of 5% (Liu et al., 2012).

Food-grade L-methionine, mainly used in human nutrition and medicine, amounts to only 5% of the whole methionine market, but offers a substantially higher margin. Biobased methionine serves the animal feed market in organic farming, in which legislation prohibits or limits the use of fossil-based feed additives (Willke, 2014).

2.2.6 Thermochemical Conversion: Fast Pyrolysis

In addition to the biochemical conversion, thermal conversion for relatively dry, woody materials is a suitable option. Currently, fast pyrolysis is a promising thermal treatment option for converting lignocellulosic biomass into liquid energy carriers or as a drop-in compound in existing refineries.

In fast pyrolysis, biomass decomposes quickly at approximately 475°C in the absence of oxygen, generating primarily condensable vapors and aerosols, as well as smaller amounts of char and noncondensable gases. After cooling and condensation, a dark brown, homogeneous, mobile liquid with a heating value about half that of conventional fuel oil is formed.

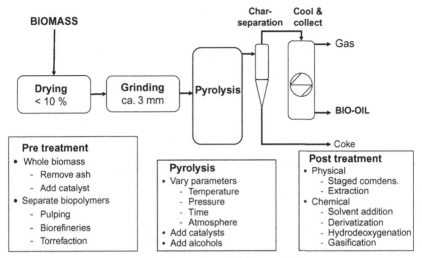

Figure 2.5 Basic process scheme of fast pyrolysis and additional process options.

The essential features of the utilization of fast pyrolysis for producing liquids are:

1. Very high heating rates and very high heat transfer rates at the biomass particle reaction interface, usually requiring a finely ground biomass feed of typically less than 3 mm, as biomass generally has a low thermal conductivity;
2. Carefully controlled pyrolysis reaction temperature of around 500 °C to maximize the liquid yield for most biomass;
3. Short hot vapor residence times of approximately 2 seconds to minimize secondary reactions;
4. Rapid and complete removal of product char to minimize cracking of vapors; and
5. Rapid cooling of the pyrolysis vapors to give the bio-oil product.

The main product, bio-oil, is obtained at yields of up to 75 wt% on a dry feed basis, together with byproduct char and noncondensable gas, which can be used within the system to provide the process heat requirements. Flue gas and ash are the only waste products. However, liquid yield depends on biomass, temperature, hot vapor residence time, char separation, and biomass ash content and composition. The latter two have a catalytic effect on vapor cracking at high temperatures.

The basic fast pyrolysis process is depicted in Fig. 2.5. Fig. 2.5 also demonstrates the versatility of fast pyrolysis processes and the most relevant process option at each process stage. The product quality and composition can be influenced through:

- pretreatment steps, such as ash removal, catalyst addition, biopolymer separation by various processes;
- changing pyrolysis conditions by varying parameters such as temperature, pressure, time, and atmosphere and addition of alcohols and catalysts;
- posttreatment, ie, both physical methods such as staged condensation and extraction and chemical procedures such as solvent addition, derivatization, hydrodeoxygenation, and gasification.

2.2.6.1 Fast Pyrolysis Reactors

Reactors serve as the heart of a fast pyrolysis process. Although it most likely represents only about 10–15% of the total capital cost of an integrated system, most research and development has focused on developing and testing varying reactor configurations on a variety of feedstocks. More recently, increased attention has been paid to control and improve liquid quality and liquid collection systems. Several comprehensive reviews of fast pyrolysis processes for bio-oil production are available (Kersten et al., 2005; Mohan et al., 2006; Bridgwater, 2009).

The main conversion reactor technologies include bubbling fluid beds, circulating fluid beds, and transported beds, the rotating cone, a type of transported bed reactor, and ablative pyrolysis, which are all described in Bridgwater (2012). The key requirements in the design and operation of a fast pyrolysis process are heat transfer and char removal, as char and ash are catalytically active (Bridgwater, 2012) and must be removed from the pyrolysis reactor as quickly as possible in order to preserve the quality of the bio-oil product.

2.2.6.2 Pyrolysis Liquid: Bio-Oil

Crude pyrolysis liquid, or bio-oil, is dark brown, viscous, and similar to biomass in elemental composition. Bio-oil is composed of a very complex mixture of oxygenated hydrocarbons with a relatively high proportion of water. It is typically unstable during storage.

Typical organic yields and their variation with temperature for a woody feedstock are shown in Fig. 2.6. Similar results are obtained for most biomass feedstocks, although the maximum yield can occur between 480°C and 520°C, depending on the feedstock. Grasses, for example, tend to give maximum bio-oil yields of around 55–60 wt% on a dry feed basis at the lower end of this temperature range.

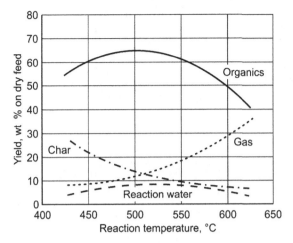

Figure 2.6 Variation of products from Aspen Poplar with temperature (Meier et al., 2013).

Bio-oil is formed by rapidly quenching, and thus "freezing," the intermediate products of flash degradation of hemicellulose, cellulose, and lignin. The bio-oil thus contains many reactive species, which contribute to its unique attributes. Bio-oil can be considered a microemulsion in which the continuous phase is an aqueous solution of holocellulose decomposition products, that stabilizes the discontinuous phase of pyrolytic lignin macromolecules through mechanisms such as hydrogen bonding. Aging or instability is believed to result from a breakdown in this microemulsion.

2.2.6.3 Bio-Oil Characteristics[4]
Fast pyrolysis liquid contains approximately 25 wt% water and has a heating value of about 17 MJ/kg. Although the fast pyrolysis liquid is widely referred to as "bio-oil," it will not mix with any hydrocarbon liquids. It is composed of a complex mixture of oxygenated compounds that provide both the potential and challenge for utilization. For any application, there are many particular characteristics of bio-oil that require consideration. These have been extensively reviewed (Bridgwater, 2011, 2012). Key properties of bio-oil are summarized in Table 2.4.

Depending on the initial feedstock and the mode of fast pyrolysis, the color can be almost black, through dark red-brown to dark green, being

---

[4]This section is based on the report of the FP7 project, DIBANET, "Report on bio-oil treatment and gas production," see www.dibanet.org/downloads.

Table 2.4 Typical properties of wood-derived crude bio-oil

Physical property	Typical value	
Water content	25%	
pH	2.5	
Specific gravity	1.2	
Elemental analysis (on dry basis)	C	56%
	H	6%
	O	38%
	N	0–0.1%
HHV (as produced)	17 MJ/kg	
Viscosity (20 °C and 25% water)	40–100 mPas	
Solids (char)	0.1%	
Vacuum distillation residue	Up to 50%	

influenced by the presence of microcarbon in the liquid and chemical composition. Hot vapor filtration gives a more translucent red-brown appearance owing to the absence of char. High nitrogen content can impart a dark green tinge to the liquid. The bio-oil contains varying quantities of water, and forms a stable single-phase mixture, ranging from about 15 wt% to an upper limit of about 30 wt% water, depending on the feed material, how it was produced, and subsequently collected.

Bio-oil can tolerate the addition of some water, but too much water will lead to phase separation. The addition of water reduces viscosity, which is useful; reduces heating value, which means that more bio-oil is required to meet a given duty; and can improve stability. The effect of water in bio-oil is therefore complex and important.

Bio-oil cannot be dissolved in water, and is miscible with polar solvents such as methanol, acetone, and is totally immiscible with petroleum-derived fuels. This is due to the high oxygen content of around 35–40 wt%, which is similar to that of biomass, and explains many of its chemical characteristics. Complex catalytic processes are necessary to remove this oxygen. Detail on this upgrading process is provided below.

The density of the bio-oil is very high at around 1.2 kg/L, compared with light fuel oil at around 0.85 kg/L. This means that the bio-oil has about 42% of the energy content of fuel oil on a weight basis, but 61% on a volumetric basis. This has implications for the design and specification of equipment such as pumps and atomizers in boilers and engines.

Viscosity is important in many fuel applications (Diebold et al., 1997). The viscosity of produced bio-oil can vary from as low as $25 \, m^2/second$ to as high as $1000\,m^2/second$ (measured at $40°C$). However, even high viscosities can result, depending on the feedstock, the water content of the bio-oil, the amount of light ends collected, and the extent to which the oil has aged.

Bio-oil cannot be completely vaporized. If the oil is heated to more than $100°C$, it rapidly reacts and eventually produces a solid residue of around $50\,wt\%$ of the original liquid, and some distillate containing volatile organic compounds and water. While bio-oil has been successfully stored for several years in normal storage conditions in polyolefin plastic drums without any deterioration that would prevent its use in any of the applications tested to date, it does change slowly with time. Most noticeably, there is a gradual increase in viscosity. More recent samples that have been distributed for testing have shown substantial improvements in consistency and stability, demonstrating the improvement in process design and control as the technology develops.

Aging is a well-known phenomenon caused by continued slow secondary reactions in the bio-oil which manifests as an increase in viscosity with time. It can be reduced or controlled by the addition of alcohols such as ethanol or methanol. It is exacerbated or accelerated by the presence of fine char (Diebold, 2002).

Approximately 300 chemical components are detectable by gas chromatography in bio-oil. They can be divided into chemical groups such as organic acids, nonaromatic carbonyls and alcohols, heterocyclic furans and pyrans, aromatics, phenols, and sugars (mainly anhydrosugars).

2.2.6.4 Applications of Bio-Oil

Bio-oil is primarily used in thermal and/or chemical applications. It can substitute for fuel oil in static applications such as boilers and turbines (Czernik and Bridgwater, 2004). Experiments have also been carried out with stationary diesel engines (Solantausta et al., 1994). Gasification of bio-oil is an ongoing activity at Karlsruhe Institute of Technology, Germany. In their bioliq process, bio-oil from straw pyrolysis is mixed with char to form a slurry which is subsequently gasified in a pressurized entrained flow gasifier. The resulting syngas is cleaned and used for the synthesis of dimethyl-ether, primarily used as transportation fuel (Dahmen et al., 2012).

Fig. 2.7 shows various use options of bio-oil. The only nonenergy application is the production of liquid smoke aroma and flavor enhancer. In the European Community, several companies have applied for product permission in the European market (Theobald et al., 2012). All other chemical applications are being studied at the laboratory or technical scale.

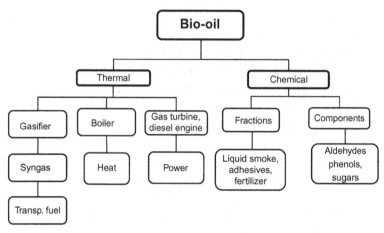

Figure 2.7 Principal pathways for bio-oil applications.

2.3 SUMMARY

Today, the use of nonfood biomass for the production of bioenergy and bio-based chemicals is preferred over the use of food crops. A wide range of nonfood biomass and conversion technologies can be used for the production of bioenergy and biobased products. The fermentation of lignocellulosic-derived sugar and the thermochemical conversion of biomass (eg, fast pyrolysis) are relevant conversion technologies. For both, the (lignocellulosic) biomass must be pretreated or at least conditioned prior to conversion.

Fast pyrolysis produces bio-oil, which can be used directly in stationary boilers. Bio-oil needs upgrading before it is suitable as fuel in transport systems such as aviation or shipping. For example, desired characteristics of jet fuels are being miscible with fossil jet fuels, remaining stable during storage, and having a suitable freeze and boiling point. Hydrocracking and hydrodeoxygenation are the two main approaches for upgrading of bio-oil to hydrocarbons that would be suitable blending agents with petroleum-derived fuels. Currently economic constraints inhibit the large-scale use of bio-oil-derived transport fuels; the same applies for bio-oil-derived materials. The use of bio-oil for material/chemical purposes is currently extremely limited.

Different technologies are involved in the initial conversion of biomass to sugars, since pretreatment is common to all lignocellulosic-sugar value chains. Technical obstacles in existing pretreatment processes for fermentation include insufficient separation of cellulose and lignin, formation of byproducts that inhibit downstream fermentation, high use of chemicals and/or energy, as well as high costs for enzymes, although the latter has decreased substantially recently. There is no single preferred pretreatment method or combination of methods, but mild acid hydrolysis and steam explosion, dilute and concentrated acid, and mild alkaline processes will be applied in the near future. Feedstock pretreatment, conversion, and product purification are linked processes and must be optimized as a whole system.

The most relevant products derived from sugars are chemicals having bifunctional groups suitable as C2–C6 building blocks. Ethanol, acetone, and butanol are also fermentation products. The catalytic conversion of intermediates, such as ethanol and glycerol, to a range of large-volume chemicals will most likely be part of the technology portfolio, particularly in regions where downstream infrastructure exists.

The integration of different pretreatment and conversion technologies in biorefineries can maximize the use of all biomass components and improve eco-efficiency of the whole value chain. In the long term, thermochemical and biochemical conversion of lignocellulosic biomass are promising technologies for the production of biofuels and biobased chemicals.

REFERENCES

Alvira, P., Tomás-Pejó, E., Ballesteros, M., Negro, M.J., 2010. Pretreatment technologies for an efficient bioethanol production process based on enzymatic hydrolysis: a review. Bioresour. Technol. 101 (13), 4851–4861.

Amores, I., Ballesteros, I., Manzanares, P., Saez, F., Michelena, G., Ballesteros, M., 2013. Ethanol Production from Sugarcane Bagasse Pretreated by Steam Explosion. Electron. J. Energy Environ. 1, 25–36.

Atsumi, S., Cann, A.F., Connor, M.R., Shen, C.R., Smith, K.M., Brynildsen, M.P., et al., 2008. Metabolic engineering of Escherichia coli for 1-butanol production. Metab. Eng. 10 (6), 305–311.

Bozell, J.J., Petersen, G.R., 2010. Technology development for the production of biobased products from biorefinery carbohydrates-the US Department of Energy's "Top 10" revisited. Green Chem. 12 (4), 539–554.

Bridgwater, A.V., 2009. Technical and Economic Assessment of Thermal Processes for Biofuels.: 19–24, NNFCC project 08/018, <http://www.globalbioenergy.org/uploads/media/0906_COPE_-_Technical_and_economic_assessment_of_thermal_processes_for_biofuels.pdf>.

Bridgwater, A.V., 2011. Upgrading biomass fast pyrolysis liquids. In: Brown, R.C. (Ed.), Thermochemical Processing of Biomass: Conversion into Fuels, Chemicals and Power Wiley-Blackwell, Chichester, UK, pp. 157–199.

Bridgwater, A.V., 2012. Review of fast pyrolysis of biomass and product upgrading. Biomass Bioenerg. 38 (0), 68–94.

Bringezu, S. 2009. Assessing Biofuels. ISBN: 978-92-807-3052-4.

Bringezu, S., O'Brien, M., Schütz, H., 2012. Beyond biofuels: assessing global land use for domestic consumption of biomass: a conceptual and empirical contribution to sustainable management of global resources. Land Use Policy 29 (1), 224–232.

Broch, A., Hoekman, S.K., Unnasch, S., 2013. A review of variability in indirect land use change assessment and modeling in biofuel policy. Environ. Sci. Policy 29 (0), 147–157.

Chiesa, S., Gnansounou, E., 2014. Use of empty fruit bunches from the oil palm for bioethanol production: a thorough comparison between dilute acid and dilute alkali pretreatment. Bioresour. Technol. 159 (0), 355–364.

Chowdhury, A., Debarati, M., Dipa, B., 2013. Biolubricant synthesis from waste cooking oil via enzymatic hydrolysis followed by chemical esterification. J. Chem. Technol. Biotechnol. 88 (1), 139–144.

Cui, L., Liu, Z., Hui, L.-F., Si, C.-L., 2011. Effect of cellobiase and surfactant supplementation on the enzymatic hydrolysis of pretreated wheat straw. BioResources 6 (4), 3836–3849.

Czernik, S., Bridgwater, A.V., 2004. Overview of applications of biomass fast pyrolysis oil. Energ. Fuels 18 (2), 590–598.

Dahmen, N., Dinjus, E., Kolb, T., Arnold, U., Leibold, H., Stahl, R., 2012. State of the art of the bioliq (R) process for synthetic biofuels production. Environ. Prog. Sustain. Energ. 31 (2), 176–181.

Dale, B.E., Kim, S., 2011. Response to comments by O'Hare et al., on the paper indirect land use change for biofuels: testing predictions and improving analytical methodologies. Biomass Bioenerg. 35 (10), 4492–4493.

Diebold, J.P., 2002. A review of the chemical and physical mechanisms of the storage stability of fast pyrolysis bio-oils In: Bridgwater, A.V. (Ed.), Fast Pyrolysis of Biomass: A Handbook, vol. 2 CPL Press, Newbury, UK, pp. 243–292.

Diebold, J.P., Milne, T.A., Czernik, S., Oasmaa, A., Bridgwater, A.V., Cuevas, A., 1997. Proposed specifications for various grades of pyrolysis oils. In: Bridgwater, A.V., Boocock, D.G.B. (Eds.), Thermochemical Biomass Conversion Blackie Academic & Professional, London.

Fargione, J., Hill, J., Tilman, D., Polasky, S., Hawthorne, P., 2008. Land clearing and the biofuel carbon debt. Science 319 (5867), 1235–1238.

Gírio, F.M., Fonseca, C., Carvalheiro, F., Duarte, L.C., Marques, S., Bogel-Lukasik, R., 2010. Hemicelluloses for fuel ethanol: a review. Bioresour. Technol. 101 (13), 4775–4800.

Gnansounou, J.D.V.E., 2008. Techno-economic and environmental evaluation of lignocellulosic biochemical refineries: need for a modular platform for integrated assessment (MPIA). J. Sci. Ind. Res. 67, 927–940.

Goh, C.S., Tan, H.T., Lee, K.T., Brosse, N., 2011. Evaluation and optimization of organosolv pretreatment using combined severity factors and response surface methodology. Biomass. Bioenerg. 35 (9), 4025–4033.

Gong, Z., Shen, H., Wang, Q., Yang, X., Xia, H., Zhao, Z., 2013. Efficient conversion of biomass into lipids by using the simultaneous saccharification and enhanced lipid production process. Biotechnol. Biofuels 6 (36).

Ha, S.-J., Galazka, J.M., Rin Kim, S., Choi, J.-H., Yang, X., Seo, J.-H., et al., 2011. Engineered Saccharomyces cerevisiae capable of simultaneous cellobiose and xylose fermentation. Proc. Natl. Acad. Sci. U S A 108 (2), 504–509.

Hendriks, A.T.W.M., Zeeman, G., 2009. Pretreatments to enhance the digestibility of lignocellulosic biomass. Bioresour. Technol. 100 (1), 10–18.

Hiloidhari, M., Das, D., Baruah, D.C., 2014. Bioenergy potential from crop residue biomass in India. Renew. Sustain. Energ. Rev. 32 (0), 504–512.

Idreesa, M., Adnan, A., Sheikhb, S., Qureshic, F.A., 2013. Optimization of dilute acid pretreatment of water hyacinth biomass for enzymatic hydrolysis and ethanol production. EXCLI J. 12, 30–40.

Jiang, D., Zhuang, D., Fu, J., Huang, Y., Wen, K., 2012. Bioenergy potential from crop residues in China: availability and distribution. Renew. Sustain. Energ. Rev. 16 (3), 1377–1382.

Kersten, S.R.A., Wang, X.Q., Prins, W., van Swaaij, W.P.M., 2005. Biomass pyrolysis in a fluidized bed reactor. Part 1: literature review and model simulations. Ind. Eng. Chem. Res. 44 (23), 8773–8785.

Kim, H., Kim, S., Dale, B.E., 2009. Biofuels, land use change, and greenhouse gas emissions: some unexplored variables. Environ. Sci. Technol. 43 (3), 961–967.

Kim, S., Dale, B.E., 2011. Indirect land use change for biofuels: testing predictions and improving analytical methodologies. Biomass Bioenerg. 35 (7), 3235–3240.

Kim, S., Dale, B.E., Ong, R.G., 2012. An alternative approach to indirect land use change: allocating greenhouse gas effects among different uses of land. Biomass Bioenerg. 46 (0), 447–452.

Kline, K.L., Oladosu, G.A., Dale, V.H., McBride, A.C., 2011. Scientific analysis is essential to assess biofuel policy effects: In response to the paper by Kim and Dale on "Indirect land-use change for biofuels: testing predictions and improving analytical methodologies". Biomass Bioenerg. 35 (10), 4488–4491.

Lai, C., Tu, M., Shi, Z., Zheng, K., Olmos, L.G., Yu, S., 2014. Contrasting effects of hardwood and softwood organosolv lignins on enzymatic hydrolysis of lignocellulose. Bioresour. Technol. 163, 320–327.

Li, Q., Lei, J., Zhang, R., Li, J., Xing, J., Gao, F., et al., 2013. Efficient decolorization and deproteinization using uniform polymer microspheres in the succinic acid biorefinery from bio-waste cotton (*Gossypium hirsutum* L.) stalks. Bioresour. Technol. 135, 604–609.

Linde, M., Jakobsson, E.L., Galbe, M., Zacchi, G., 2008. Steam pretreatment of dilute H2SO4-impregnated wheat straw and SSF with low yeast and enzyme loadings for bioethanol production. Biomass Bioenerg. 32 (4), 326–332.

Liu, Z., Ying, Y., Li, F., Ma, C., Xu, P., 2010. Butanol production by Clostridium beijerinckii ATCC 55025 from wheat bran. J. Ind. Microbiol. Biotechnol. 37 (5), 495–501.

Liu, L., Ye, X.P., Bozell, J.J., 2012. A comparative review of petroleum-based and bio-based acrolein production. ChemSusChem 5 (7), 1162–1180.

Liu, Y., 2000. New Biodegradable Polymers From Renewable Resources. PhD, Royal Institute of Technology, Stockholm, Sweden.

Luo, J., Xia, L., Lin, J., Cen, P., 1997. Kinetics of Simultaneous Saccharification and Lactic Acid Fermentation Processes. Biotechnol. Prog. 13 (6), 762–767.

Luo, Y., Stichnothe, H., Schuchardt, F., Li, G., Huaitalla, R.M., Xu, W., 2013. Life cycle assessment of manure management and nutrient recycling from a Chinese pig farm. Waste Manag. Res. 32, 4–12.

Mcbride, J.E. (2010). Yeast expressing cellulases for simultaneous saccharification and fermentation using cellulose. S.L. Mascoma Corporation [US/US]; 67 Etna Road, New Hampshire 03766 (US). WO/2010/060056.

Meier, D., van de Beld, B., Bridgwater, A.V., Elliott, D.C., Oasmaa, A., Preto, F., 2013. State-of-the-art of fast pyrolysis in IEA bioenergy member countries. Renew. Sustain. Energ. Rev. 20, 619–641.

Mergner, R., Janssen, R., Rutz, D., De Bari, I., Sissot, F., 2013. Lignocellulosic ethanol process and demonstration. A handbook Part I. WIP Renewable Energies, Munich, Germany.

Mohan, D., Pittman, C.U., Steele, P.H., 2006. Pyrolysis of wood/biomass for bio-oil: a critical review. Energ. Fuels 20 (3), 848–889.

Metzger, J.O., 2006. Production of Liquid Hydrocarbons from Biomass. Angewandte Chemie International Edition 45 (5), 696–698.

Montanarella, L., Vargas, R., 2012. Global governance of soil resources as a necessary condition for sustainable development. Curr. Opin. Environ. Sustain. 4 (5), 559–564.

Mosier, N., Wyman, C., Dale, B., Elander, R., Lee, Y.Y., Holtzapple, M., et al., 2005. Features of promising technologies for pretreatment of lignocellulosic biomass. Bioresour. Technol. 96 (6), 673–686.

O'Hare, M., Delucchi, M., Edwards, R., Fritsche, U., Gibbs, H., Hertel, T., et al., 2011. Comment on "Indirect land use change for biofuels: Testing predictions and improving analytical methodologies" by Kim and Dale: statistical reliability and the definition of the indirect land use change (iLUC) issue. Biomass. Bioenerg. 35 (10), 4485–4487.

Panagiotopoulos, I.A., Lignos, G.D., Bakker, R.R., Koukios, E.G., 2012. Effect of low severity dilute-acid pretreatment of barley straw and decreased enzyme loading hydrolysis on the production of fermentable substrates and the release of inhibitory compounds. J. Clean. Prod. 32 (0), 45–51.

Papa, G., Rodriguez, S., George, A., Schievano, A., Orzi, V., Sale, K.L., et al., 2015. Comparison of different pretreatments for the production of bioethanol and biomethane from corn stover and switchgrass. Bioresour. Technol. 183, 101–110.

Posada, J.A., Patel, A.D., Roes, A., Blok, K., Faaij, A.P.C., Patel, M.K., 2013. Potential of bioethanol as a chemical building block for biorefineries: Preliminary sustainability assessment of 12 bioethanol-based products. Bioresour. Technol. 135 (0), 490–499.

Qureshi, N., Meagher, M.M., Huang, J., Hutkins, R.W., 2001. Acetone butanol ethanol (ABE) recovery by pervaporation using silicalite–silicone composite membrane from fed-batch reactor of Clostridium acetobutylicum. J. Memb. Sci. 187 (1–2), 93–102.

Qureshi, N., Saha, B.C., Hector, R.E., Hughes, S.R., Cotta, M.A., 2008. Butanol production from wheat straw by simultaneous saccharification and fermentation using Clostridium beijerinckii: Part I—batch fermentation. Biomass. Bioenerg. 32 (2), 168–175.

Reisinger, M., Tirpanalan, Ö., Huber, F., Kneifel, W., Novalin, S., 2014. Investigations on a wheat bran biorefinery involving organosolv fractionation and enzymatic treatment. Bioresour. Technol. 170, 53–61.

Santosa, S.J., 2008. Palm oil boom in Indonesia. Clean 36 (5–6), 453–465.

Schmidhuber, J. (2007). Biofuels: an emerging threat to Europe's Food Security? Notre Europe.

Searchinger, T., Heimlich, R., Houghton, R.A., Dong, F., Elobeid, A., Fabiosa, J., et al., 2008. Use of U.S. croplands for biofuels increases greenhouse gases through emissions from land-use change. Science 319 (5867), 1238–1240.

Searle, S.Y., Malins, C.J., 2014. Will energy crop yields meet expectations? Biomass. Bioenerg. 65 (0), 3–12.

Shi, J., Ebrik, M.A., Wyman, C.E., 2011. Sugar yields from dilute sulfuric acid and sulfur dioxide pretreatments and subsequent enzymatic hydrolysis of switchgrass. Bioresour. Technol. 102 (19), 8930–8938.

Silveira, M., Jonas, R., 2002. The biotechnological production of sorbitol. Appl. Microbiol. Biotechnol. 59 (4), 400–408.

Solantausta, Y., Nylund, N.O., Gust, S., 1994. Use of pyrolysis oil in a test diesel-engine to study the feasibility of a diesel power-plant concept. Biomass Bioenerg. 7 (1–6), 297–306.

Theobald, A., Arcella, D., Carere, A., Croera, C., Engel, K.H., Gott, D., et al., 2012. Safety assessment of smoke flavouring primary products by the European Food Safety Authority. Trends Food Sci. Technol. 27 (2), 97–108.

US-DOE, 2011. In: Perlack, R.D., Stokes, B.J. (Eds.), U.S. Billion-Ton Update: Biomass Supply for a Bioenergy and Bioproducts Industry U.S. Department of Energy, Oak Ridge National Laboratory, Oak Ridge, pp. 227.

van Zyl, W., den Haan, R., la Grange, D., 2013. Developing cellulolytic organisms for consolidated bioprocessing of lignocellulosics. In: Gupta, V.K., Tuohy, M.G. (Eds.), Biofuel Technologies Springer Berlin Heidelberg, pp. 189–220.

Vidal Jr., B., Dien, B., Ting, K.C., Singh, V., 2011. Influence of feedstock particle size on lignocellulose conversion—a review. Appl. Biochem. Biotechnol. 164 (8), 1405–1421.

Vlysidis, A., Binns, M., Webb, C., Theodoropoulos, C., 2011. A techno-economic analysis of biodiesel biorefineries: assessment of integrated designs for the co-production of fuels and chemicals. Energy 36 (8), 4671–4683.

Werpy, T., Petersen, G., Aden, A., Bozell, J., Holladay, J., White, J., et al., 2004. Top Value Added Chemicals from Biomass Volume I—Results of Screening for Potential Candidates from Sugars and Synthesis Gas. U.S. DoE, Springfield, US.

Willke, T., 2014. Methionine production—a critical review. Appl. Microbiol. Biotechnol. 98 (24), 9893–9914.

Yang, B., Wyman, C.E., 2008. Pretreatment: the key to unlocking low-cost cellulosic ethanol. Biofuels, Bioprod. Biorefining 2 (1), 26–40.

Zhu, F., Cai, J., Wu, X., Huang, J., Huang, L., Zhu, J., et al., 2013. The main byproducts and metabolic flux profiling of γ-PGA-producing strain B. subtilis ZJU-7 under different pH values. J. Biotechnol. 164 (1), 67–74.

Zhuang, J., Liu, Y., Wu, Z., 2009. Hydrolysis of wheat straw hemicellulose and detoxification of hydrolysate for xylitol production. BioResources 4 (2), 674–686.

CHAPTER 3

Biorefineries: Industry Status and Economics

H. Stichnothe[1], D. Meier[2] and I. de Bari[3]
[1]Thünen Institute of Agricultural Technology, Braunschweig, Germany
[2]Thünen Institute of Wood Research, Hamburg, Germany
[3]Division of Bioenergy, Biorefinery and Green Chemistry, ENEA Centro Ricerche Trisaia, Policoro, Italy

Contents

Abstract

This chapter focuses on techno-economic assessments of biobased products from fermentation and fast pyrolysis. We briefly describe underlying methodologies and summarizea number of literature studies dealing with conversion pathways using fermentation and fast pyrolysis. However, data are limited on the techno-economic evaluation of innovative biorefinery processes. We identify methodological challenges for techno-economic assessment of biofuels and biobased chemicals and describe limits of comparing results from different studies. We depict the lessons learned from lignocellulosic ethanol and show examples of biobased chemicals and bio-oil that are already produced at large scale. In the longterm, biobased chemicals and biofuels must compete on cost and performance with petrochemicals and petroleum fuels, however at present biofuels and biobased chemicals derived from lignocellulosic biomass are hardly cost-competitive.

Developing the Global Bioeconomy.
DOI: http://dx.doi.org/10.1016/B978-0-12-805165-8.00003-3

© 2016 Elsevier Inc.
All rights reserved.

3.1 INTRODUCTION

Frequently, feedstock composition determines the choice of the conversion strategy, and a range of technologies are available for bioenergy production. The use of versatile, robust technologies is one of the key factors in biorefineries. The synergetic combination of process technologies can lead to the development of advanced biorefineries where nonfood biomass, preferably from residues and waste, is converted by a combination of mechanical, thermochemical, chemical, and biochemical processes, into a range of materials, chemicals, and energy. Hence, the maximum value is achieved from each feedstock.

In fact, the possibility of diversifying both the feedstock and the final products would help biorefineries cope with continuously changing market dynamics and connected commercialization constraints, because the same infrastructure could be leveraged to produce a broad range of products in response to changing market conditions.

Currently, biofuel production is driven by politically motivated incentive schemes as well as the desire for a more sustainable fuel alternative. Although there is already a market for a number of biobased plastics and other biobased chemicals, the demand is considerably lower in comparison to traditional fossil fuel-based plastics. However, the market for biobased chemicals is anticipated to increase significantly in the near future, for example, the global biobased biodegradable plastics market is forecast to grow by 18% between 2014 and 2020 according to a report of Future Market Insight.[1] Integrated biorefineries producing biobased products using bioenergy produced on-site and exporting just the bioenergy surplus is a promising way forward.

At present the majority of biorefineries do not provide the synergetic combination. Most large-scale fermentation plants still run on starch or sugar as feedstock, but there is a preference to use lignocellulosic-derived sugars in the future, because of rising sugar prices and public concerns. Large-scale fast pyrolysis plants produce mainly bio-oil for stationary burners and in the midterm bio-oils could be upgraded to transportation fuels and platform chemicals. Although promising, integrated biorefineries still have to prove that they can become cost-competitive.

---

[1] http://www.prweb.com/releases/biodegradable-packaging/market-2019-forecasts/prweb11796468.htm

3.2 ECONOMICS

Cost estimation is a specialized subject and a profession in its own right (Towler, 2007). The development of a new biorefinery, its design and construction, requires huge investments; cost estimations are often paramount for deciding the economic viability of biorefineries, and must be performed on a case-by-case basis. However, it is possible to make (rough) cost estimations based on data from demo plants, process modeling, and/or literature at various stages of the biorefinery development.

3.2.1 Economic Considerations

Total cost can be divided into capital expenditures (CAPEX) and operating expenditure (OPEX). CAPEX can be subdivided into plant costs, off-site costs, and engineering costs. Plant costs represent the capital necessary for the installed process equipment with all auxiliaries that are needed for complete process operation (ie, piping, instrumentation, insulation, foundations, and site preparation). Off-site costs are not directly related to the process operation, they rather include costs of the addition to the site infrastructure, for example, power generation units, boilers, pipelines, offices, etc. According to the American Association of Cost Engineers, fixed capital investment can be divided into outside battery limits (OBL) or off-site and inside battery limits (IBL).IBL comprises one or more geographic boundaries, to specify the area where production takes place, enclosing all associated equipment and production facilities. OBL includes facilities such as storage, utilities, administration buildings, or auxiliary facilities. Engineering costs include costs for detailed design of equipment, construction supervision, project management, etc.

OPEX consists of fixed and variable costs. Variable costs comprise cost of feedstock and supplies, waste management, product packaging, finished and semifinished products in stock, etc. Fixed costs comprise salaries, taxes, license fees, interest payments, marketing costs, etc.

A generic approach on how to conduct the economic assessment, the most relevant cost parameters, and ballpark figures for first-generation (1G) feedstock is depicted in Fig. 3.1.

Investment costs are an important contributor to the economic feasibility of lignocellulosic biorefineries (Anex et al., 2010; Brown, 2015). Investment costs are more important for second-generation (2G) biorefineries than for 1G biorefineries because feedstock costs are lower and processing lignocellulosic biomass is more challenging

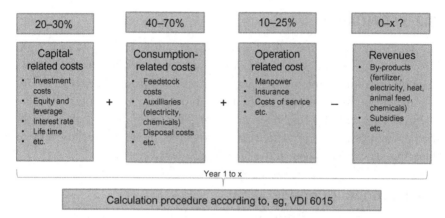

Figure 3.1 Model for cost calculation of biobased products. *Modified from Zinoviev, S., Müller-Langer, F., Das, P., Bertero, N., Fornasiero, P., Kaltschmitt, M., et al., 2010. Next-generation biofuels: survey of emerging technologies and sustainability issues. ChemSusChem 3(10), 1106–1133 and Posada, J.A., Patel, A.D., Roes, A., Blok, K., Faaij, A.P.C., Patel, M.K., 2013. Potential of bioethanol as a chemical building block for biorefineries: preliminary sustainability assessment of 12 bioethanol-based products. Bioresour. Technol. 135, 490–499.*

(Hytönen, 2009; Zinoviev et al., 2010; Brown, 2015); hence capital investment costs are higher.

Innovative new conversion technologies usually follow a development pathway from the lab, to piloting, then demonstration, and finally the construction of a commercial plant. The number of years for a biobased product to reach commercialization depends heavily on economics and hence on drop-in versus nondrop-in (existing demand and infrastructure), type of conversion technology, and supply chain integration. Various methods are applied for estimating investment costs when plants or processes are scaled-up.

3.2.1.1 Factorial Method

Preliminary cost estimates can be based on the purchased equipment cost with all additional cost components based on it and estimated through different "factors," that is, certain percentages of the purchased equipment cost. The factorial cost estimation method was developed by Lang (1947, 1948) and frequently used in chemical engineering in early design stages. Although biorefineries may differ from chemical plants, a similar approach is appropriate (Stanil, 2011). The accuracy of the result of the factorial method depends on what stage the design has reached at the time the estimate is made, and on the reliability of the data available on equipment costs. Quantified predictions can only be done for a limited amount of time ahead, since uncertain market conditions and rapid technology development result in sharp price inflation.

The equation for the cost estimation is:

$$C_c = f_L * C_e,$$
(3.1)

where C_c = fixed capital costs, f_L = Lang factor, can be obtained from chemical engineers design books (eg, Peters et al., 2002; Towler, 2007), and C_e = total costs of major equipment.

The "Lang factor" depends on the type of process, that is, solids processing (f_L = 3.1), liquid processing (f_L = 4.2) and liquids–solids processing (f_L = 3.6).[2]

If more detailed information is available, compounds such as piping, electrical power, control instruments, etc., can be considered individually. The individual costs are summed to generate the total capital investment costs.

3.2.1.2 Step Accounting Method

Step accounting methods attempt to correlate the capital investment cost of a process with the values of its fundamental process parameters. These values might be the number of significant process steps, plant capacity, temperature, pressure, materials of construction, number of significant process steps, etc. The term significant process step (or functional unit) defines all equipment and auxiliaries necessary to complete an operation in the production stream, as defined by the American Association of Cost Engineers. This method is useful when the plant has a limited number of products, for example, power or fuel production, but difficult to apply for multioutput systems such as new and innovative biorefineries.

3.2.1.3 Exponential Method

The cost of a future plant can be estimated based on historical cost data of similar plants. The same applies for estimating the costs of specific process equipment, for example, heat exchanger, distillation units, etc.

$$\frac{ACC_1}{ACC_2} = \left(\frac{Q_1}{Q_2}\right)^m * corr_F,$$
(3.2)

[2] https://chemicalprojects.wordpress.com/2014/05/09/the-factorial-method-of-cost-estimation

where ACC_1 = annual capital cost of unit or scale (Q_1), for example, large scale; ACC_2 = annual capital cost of unit or scale (Q_2), for example, demo scale; m = scale exponent; and $corr_F$ = correction factor, if the design is highly comparable $corr_F = 1$.

Calculating the capital costs for a biorefinery or specific equipment is not straightforward because comparable facilities are not built yet and thus appropriate factors, for example, Lang factors, for future biorefineries are hardly available. If future biorefineries are comparable to established chemical plants, then the cost estimation methods described above can be also used to estimate the product price depending on the scale-up. They can also be used to define optimization targets for the process design and the scale-up process. An example of how the exponential method can be used is shown below.

Example:

For the scale-up the residence time and the conversion rates (cf_{1-n}) are the most relevant parameters. As first approximation product costs (PC) can be estimated based on throughput (TP), conversion factor (cf), annual capital costs (ACC), operation costs (OC), and revenue (R) according to the following equation:

$$PC = \frac{1}{TP \star cf} \star (OC + ACC) - R \tag{3.3}$$

The selling price (SP) can be expressed as function of the residence time if the throughput and the conversion rate are assumed to be constant.

$$PC(rt) = \frac{1}{TP \star cf} \star (OC(rt) + ACC(rt)) - R(rt) \leq SP \tag{3.4}$$

The acceptable residence time (rt) for a plant at technical scale can be calculated using the SP as a constraint. The revenue from byproducts depends linearly on the throughput. Hence the revenue can be expressed as function of the throughput when the SP for each byproduct (n) is used:

$$R = \sum_{1}^{n} TP \star cf_n \star SP_n \tag{3.5}$$

ACC does not increase linearly with scale; therefore the acceptable residence is higher at large scale than at technical scale for a given minimum selling price (MSP). The exponential method can be used to estimate the ACC of large-scale plants (LS) based on ACC of technical scale plants (TS). The exponential method is frequently used in chemical engineering and is expressed as:

$$ACC_{LS} = ACC_{TS} \star \left(\frac{LS}{TS}\right)^{m} \qquad (3.6)$$

where the exponent (m) usually varies between 0.6 and 0.7.

Using Eqs. (3.5) and (3.6) in Eq. (3.4) allows calculating of the acceptable residence time for a given scale-up (TS→LS). The equations above also allow defining optimization targets, for example, conversion factors for the main and the byproduct(s) under consideration of the MSP of the product(s) for a given process design. However, those calculations are based on point estimates and incur a substantial uncertainty due to price fluctuations. One method to reduce the uncertainty is to model the statistical distribution of prices in combination with sensitivity analysis methods such as Monte Carlo simulation.

The methods described above are applicable if comparable plants exist. In the last decade many plants have been built for the production of biofuels and bioenergy, therefore reliable cost data exist for those technologies. Table 3.1 provides an overview of typical investment and operation costs of matured technologies at different scales.

In general, the first-of-its-kind commercial plant shows disadvantages with regard to the total capital investment but also operational costs in comparison to succeeding installations. Several installations are frequently needed to enhance efficiencies due to technical learning. High-value material utilization of all product streams is essential to achieve profitability of complex biorefineries.

The production of bioethanol is the most matured biotechnological process. Historical data allow estimating the cost reduction potential due to technological progress, which has been made for ethanol production from woody material and agricultural residues. The techno-economic development of ethanol from cellulosic feedstock is depicted below.

In the early technology development stage ethanol production costs can be used as a proxy for other products derived from cellulosic sugars,

Table 3.1 Overview of investment and operation costs of bioenergy-related technologies[a]

Technology	Plant production capacity[b]			Investment costs (range)	Operation costs (range)	Preferred feedstock
	Minimum capacity	Typical capacity	Maximum Capacity			
Pyrolysis (fast)	15,000 t/year	45,000 t/year	300,000 t/year	600–3000 € per kW$_{inst}$	11% of investment	Wood and dry biomass
Anaerobic digestion	0.3 MW$_e$	0.5 MW$_e$	4 MW$_e$	1800–4000 € per kW$_{inst}$ (0.5–2 MW)	11% of investment	Nearly all kinds of organic biomass
First-generation bioethanol	10,000 t/year	250,000 t/year	1,000,000 t/year	700–1000 € per t/year	11% of investment	Starch (wheat, potatoes, etc.)
Esterification (biodiesel)	10,000 t/year	100,000 t/year	500,000 t/year	400–600 € per t/year	11% of investment	Oils and fats
Direct liquefaction	5000 t/year	50,000 t/year	200,000 t/year	900 € per t/year	11% of investment	All kinds of organic waste streams

[a]This information is kindly provided by NOVIS GmbH.
[b]1 t = 1000 kg = 1 Megagram (Mg).

for example, butanol, furan derivatives, etc. However, that is just a rough estimation because equipment costs and energy costs differ, although similar equipment is needed for purification and upgrading.

3.2.2 Economic Lessons Learned From Bioethanol and Bio-Oil Derived From Lignocellulosic Biomass

The National Renewable Energy Laboratory (NREL), Golden, CO, USA, has developed case models that document progress and cost targets for energy carrier production from cellulosic feedstock (Anex et al., 2010; Kazi et al., 2010; Swanson et al., 2010; Wright et al., 2010; Davis et al., 2013). The economics analysis includes at a conceptual level of process a design to develop flow diagrams using commercial simulation tools, for example, ASPEN PLUS. The capital investment costs are estimated based on the plant design using the factorial method; working costs are estimated based on similar chemical plants (Wright et al., 2010). The simulation software is also used to calculate the mass and energy flows in order to estimate the operation costs. Detailed information of the technology model is available (Aden, 2008, 2009).

The NREL cost model for ethanol and/or hydrocarbons from cellulosic feedstock is based on a plant capacity of 2000 t (biomass demand) per day (t/day). For bioethanol the model includes the following process steps:

- Pretreatment (dilute acid and enzymatic digestion);
- Biological conversion of sugars to ethanol;
- Product recovery and upgrading (assuming lignin for combustion);
- Waste water treatment.

Kumar and Murthy (2011) have modeled the conversion of grass straw to ethanol; grass straw is a byproduct of grass seed production and has a cellulose content of 31%. They used dilute acid, dilute alkali, hot water, and steam explosion as pretreatment in their model to estimate the production costs of all pretreatment methods followed by fermentation. Lignin residue was sufficient to cover the energy demand of pretreatment, conversion, and upgrading. The authors estimated ethanol production costs as US$ 1.06, 1.12, 1.02, and 1.09 per kg ethanol for dilute acid, dilute alkali, hot water, and steam explosion pretreatment, respectively, at a plant capacity of approximately 750 t/day (Kumar and Murthy, 2011). The main factors were feedstock costs and enzyme costs. Kazi et al. (2010) investigated the conversion of corn stover to ethanol and came to the same conclusion that enzyme costs and feedstock costs are the most important factors.

There are several optimization options that are suitable to reduce the production costs, for example, improved conversion efficiency of underutilized biomass fractions (mainly hemicellulose) with production of additional value-added products. Besides feedstock and enzyme costs, the pentose conversion rate is another important option for reducing ethanol production costs. Xylose is the most important pentose obtained via acid hydrolysis and can be thermochemically converted into furfural. Furfural, a common industrial chemical, is a potential platform chemical for biopolymers (Choudhary et al., 2012).

Interest in drop-in biofuels, especially for aviation, continues to grow. Tao has conducted a techno-economic assessment for the production of *n*-butanol, iso-butanol, and ethanol from corn stover. In contrast to cellulosic ethanol fermentation technology, *n*-butanol and especially iso-butanol fermentation technology, has not yet been fully demonstrated even in bench studies (Tao et al., 2014). However, the authors concluded that if fermentation of xylose and glucose to butanol would reach 85% yield, the minimum SP would then be comparable to that of ethanol.

Results from various NREL techno-economic studies about lignocellulosic bioethanol production in the period from 2001 to 2012, shown in Fig. 3.2, reveal that cost reduction up to a factor of four to five could be achieved due to technical learning over such a long time period. That may

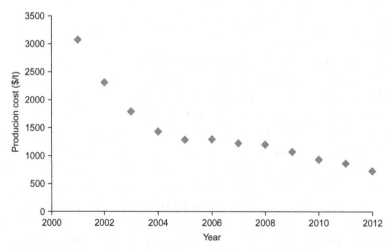

Figure 3.2 Production costs of ethanol from cellulosic feedstock (Davis et al., 2013).

also be achievable for the production of higher value-added products in the next decade.

For the production of cellulosic ethanol, cost reduction was mainly achieved through technological improvements in the areas of pretreatment and conditioning, enzymatic hydrolysis (and associated improvements in enzyme performance reflecting commercial enzyme package improvements), and fermentation, while the feedstock cost contributions to ethanol SP remained relatively constant.

For 1G biofuels, feedstock significantly influences production costs, whereas for 2G biofuels, the share decreases and becomes less than 40% (Hamelinck et al., 2005). The production cost of ethanol from cellulosic biomass is sensitive to economy of scale (Argo et al., 2013; Muth et al., 2014). However, the optimal size of biorefineries depends also upon the nature of the feedstock. Preprocessed feedstock in depot systems enables larger biorefinery sizes—at least conceptually but might compromise greenhouse gas emission (GHG)-reduction targets (Muth et al., 2014).

For processes dealing with high volumes of raw material and high capital costs, marginal changes in feedstock cost (including transport) can make the difference. Therefore, in assessing the economic viability of a lignocellulose biorefinery, trade-off between plant size and feedstock delivery costs must be taken into account. Ringer et al. (2006) provided scale-dependeny cost information on the production of bio-oil from wood chips.

Fig. 3.3 shows that the specific production costs decrease with scale, although the feedstock costs increases mainly due to transport costs. There is obviously a limit, where cost reduction due to technological progress is compensated by increasing cellulosic feedstock (and transport) cost (Argo et al., 2013). Transport costs can vary substantially. Estimation of feedstock cost is not straightforward due to the lack of formal markets for a large part of possible feedstock and due to site-specific availability and procurement constraints. This aspect is discussed in detail in chapters "Biomass Supply and Trade Opportunities of Preprocessed Biomass for Power Generation" and "Commodity Scale Biomass Trade and Integration With Other Supply Chains."

Currently, the production of hydrocarbons from cellulosic feedstock is not cost-competitive to hydrocarbons from crude oil. Davis et al. (2013) summarize their findings as: "Tailoring the hydrolysate stream to the microorganism tolerance will be essential for improving overall yields and lowering production costs. Reduction of hydrolysate conditioning

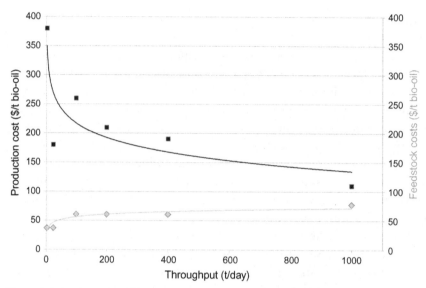

Figure 3.3 Production and feedstock costs for bio-oil from wood chips.

costs could also be realized through a better understanding of the tolerance of hydrocarbon-producing microbes to lignin and other cellulosic
sugar substrate impurities and other potential inhibitors. There are several
optimization possibilities that are suitable to reduce the production costs,
for example, improving conversion efficiency of underutilised fractions
of the biomass (mainly hemicellulose) or conversion of lignin to higher
value-added products." However, the latter would require the replacement of lignin by other energy carriers and that might lead to a trade-off
between maximizing the GHG savings and reducing the production costs.
Currently, lignin is mostly used as fuel on-site to cover the heat and/or
power demand of the biorefinery.

For the techno–economic evaluation of innovative industrial processes,
often limited data are available. The economic assessment of biorefineries depends on a number of factors such as feedstock costs (Argo et al.,
2013; Muth et al., 2014), specific preprocessing and process equipment
(Muth et al., 2014), plant capacity, etc. Moreover, some of those factors
are site-specific, for example, feedstock logistics, process waste utilization, integration in existing installations, etc. Economic estimations have
been conducted for green biorefineries in Ireland (O'Keeffe et al., 2011,
2012) and elsewhere (Höltinger et al., 2014), sugar cane biorefineries in
Brazil (Cavalett et al., 2012), for thermochemical production of biofuels
(Bridgwater, 2009; Reyes Valle et al., 2015), the production of advanced

biofuels (Turley et al., 2013), lignocellulosic biorefineries (Kazi et al., 2010; Benali et al., 2014; Cheali et al., 2015), integrating bioethanol production into Kraft pulp and paper mills (Hytönen, 2009) but also for biorefinery-relevant separation techniques (Sievers et al., 2014) and pretreatment techniques (García et al., 2011). The results of the studies are hardly comparable because assumptions such as interest rate, equipment lifetime, and modeling approaches are different.

Investment cost estimations are mostly point estimates and do not take the uncertainties inherent in the methodology into account. The accuracy of the result depends on the quality of the input data and the calculation approach. Brown (2015) has recently reviewed techno-economic studies of thermochemical cellulosic biorefinery and conducted a sensitivity analysis based on economic data available in the public domain. He concluded that the choice of estimation methodology differs across the techno-economic assessments and has a substantial impact on the final result. The difference of capital investment costs of a biorefinery with a capacity of 2000 t/day can be as high as US\$ 300,000 million depending on the chosen estimation method and assumptions (Brown, 2015). Therefore, it is important to assess the uncertainty by sensitivity analysis if estimation methods are used for supporting decision making.

Daugaard et al. (2015) discuss how increasing biorefinery capital and feedstock learning rates could significantly reduce optimal size and production costs of biorefineries. The authors suggest that there is an economic incentive to invest in strategies that increase the learning rate for producing advanced biofuels.

3.3 DEMONSTRATION AND FULL-SCALE PLANTS

The large-scale production of biobased products via fermentation of lignocellulosic-derived sugars is still in its infancy; except for bioethanol. Thermochemical conversion of lignocellulosic biomass at a largescale is more targeted to produce biofuels than biobased chemicals. Fast pyrolysis is probably the most interesting thermochemical conversion technology because the yield of bio-oil is high and it can be used both for energy and/or chemicals production.

3.3.1 Fast Pyrolysis: Current Status

Currently, the main objective of fast pyrolysis is to produce bio-oil that can be directly used in stationary boilers or in the near future after hydrogenation as transportation fuel. In the mid-term bio-oil could also be used as drop-in (after partial upgrading) in naphtha-crackers of conventional refineries or even as source for chemicals. The technological development in the area of fast pyrolysis of mainly woody biomass is driven by the demand for 2G biofuels.

3.3.1.1 Ensyn Corp (ENSYN, 2014)

Ensyn Corp (Canada) is executing its renewable fuels business in alliance with UOP, a Honeywell company. This alliance takes place through Envergent Technologies, LLC (Envergent), a joint venture between Ensyn and UOP.

A pyrolysis oil demonstration project at Manitoba Hydro involved the production and use of pyrolysis oil as a replacement for heavy fuel at the Tolko Kraft Paper Mill in The Pas, Manitoba. The equipment and services are provided by Ensyn Technologies Ltd. The cofiring demonstration plant successfully fired over 60,000 L of bio-oil in 2010 with stable combustion. Extensive emissions monitoring indicated improvements in boiler performance. A second demonstration is planned for a grain-drying operation where bio-oil will replace propane (Manitoba, 2014).

Recently, Memorial Hospital in North Conway, New Hampshire, shifted from petroleum to Ensyn's Renewable Fuel Oil (RFO) for their heating system. RFO is certified according to ASTM D7544-10 (Standard Specification for Pyrolysis Liquid Biofuel), which ensures the highest levels of quality, reliability, and performance when used as a fuel in industrial burners.

Ensyn is supplying Memorial Hospital with 300,000 gallons/year of Ensyn's RFO cellulosic biofuel under a 5-year, renewable contract. This contract allows the hospital to fully replace its petroleum fuels with Ensyn's renewable fuel, and reduce its GHG emissions from heating by approximately 85%. Memorial Hospital's boiler has been operating successfully on 100% RFO since September 2014.

Lastly, in early October 2014, Ensyn announced a 7-year renewable contract to supply Valley Regional Hospital in Claremont, New Hampshire, with 250,000 gallons/year of RFO. Under this contract, the hospital intends to replace its entire heating fuel requirements with Ensyn's RFO. Deliveries are expected to begin in April, 2015.

Ensyn also provided Petrobas with bio-oil from spruce for test trials at demo-scale. Bio-oil was directly coprocessed in a fluid catalytic cracking (FCC) unit without any type of hydrodeoxygenation and with a regular gasoil FCC feed up to 20%wt. The bio-oil and the conventional gasoil were cracked into valuable products such as LPG and gasoline (Pinho et al., 2015).

3.3.1.2 KIT (Bioliq)[3]

The bioliq process, developed at the Karlsruhe Institute of Technology (KIT), aims to produce synthetic fuels and chemicals from biomass (Dahmen et al., 2012). This process requires a feed that can easily be fed to the gasifier at elevated pressures being atomized by oxygen as the gasification agent. Fast pyrolysis was chosen as the most promising process to produce this feed, by mixing pyrolysis condensates and char to a so-called bioliqSyncrude, exhibiting a sufficiently high heating value, and being suitable for transport, storage, and processing. This slurry-gasification concept has been extended to a complete process chain via an on-site pilot plant erected KIT.

The three subsequently constructed parts of the plant, funded by the German Ministry of Food, Agriculture, and Consumer Protection, consist of the fast pyrolysis and biosyncrude production, the $5\,MW_{th}$ high-pressure entrained-flow gasifier operated at up to 8 MPa (both in cooperation with Lurgi GmbH, Frankfurt), as well as the hot gas cleaning (MUT Advanced Heating GmbH, Jena), dimethylether, and final gasoline synthesis (Chemieanlagenbau Chemnitz GmbH).

The technical concept of the pilot plant is based on the Lurgi-Ruhrgas mixer reactor, previously devoted to coal degassing or heavy crude coking. The actual flow scheme is depicted in Fig. 3.4. In November 2014, the whole process chain was demonstrated successfully at KIT and the first synthetic fuel was produced.

3.3.1.3 Fortum

The new technology has been developed into a commercial-scale concept in cooperation between Fortum, Metso, UPM, and VTT as part of TEKES Biorefine research program. Fortum commissioned a first-of-its-kind fast pyrolysis plant in 2013, which has been operating since 2014. It was the first-of-its-kind at commercial scale and is shown in Fig. 3.5. The bio-oil plant is integrated with existing combined heat and power production and an urban district heating network in Joensuu, Finland. Wood chips from forest residues are dried and then fine-milled before they are fed into the pyrolysis reactor together with hot sand. The bio-oil production capacity

[3] Partly taken from IEA Task 42 Newsletter June 2010.

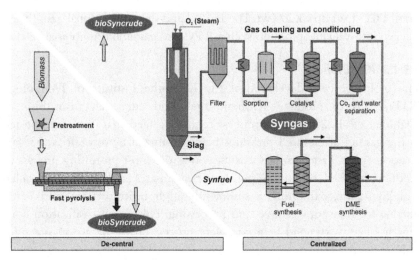

Figure 3.4 Process flow chart of the bioliq process (Bioliq, 2014).

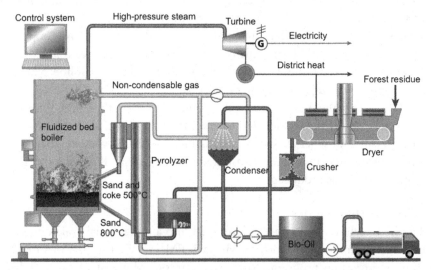

Figure 3.5 Industrial-scale integrated bio-oil plant in Joensuu, Finland (Meier et al., 2013).

is 50,000 t/year. The bio-oil and the char are conveyed to the CHP unit, where the bio-oil replaces heavy fuel oil. The steam is used the district heating network and surplus electricity is fed into the national grid. This helps to reduce CO_2 emissions by 59,000 t/year and SO_2 emissions by 320 t/year.

3.3.1.4 BTG BioLiquids BV

BTG BioLiquids BV (BTG-BTL) was established to commercialize the fast pyrolysis technology developed by BTG. The business model is to operate as a technology supplier, supplying the core components of the pyrolysis unit. A demonstration unit on a commercial scale was required to prove both the technology and the complete pyrolysis chain.

The core conversion process is a flash pyrolysis process based on BTG technology with a rotating cone. The feedstock will be crushed pellets imported from the USA and Canada via the port of Rotterdam. Excess heat will be converted into process steam to drive a steam turbine for electricity generation. Part of the low-pressure steam will be used to dry the biomass, while excess steam will be sent to AkzoNobel. The oil produced will be cofired in a novel modified gas burner at Royal Friesland Campina NV, a global dairy company. Pyrolysis off-take agreement concluded for a period of 12 years, equivalent to approximately 200 million liters (EMPYRO, 2014).

3.3.1.5 EMPYRO

Energy and Materials from Pyrolysis (EMPYRO) uses BTG fast pyrolysis technology. In May 2015 a fast pyrolysis plant started operating in Hengelo, Netherlands. In the plant, 5 t/hour of clean wood will be converted into about 3.2 t/hour of bio-oil. Excess heat generated from the combustion of the byproducts (gas and char) is used to provide heat for the biomass dryer, and to run a steam turbine to generate electricity. EMPYRO received an NTA8080 certificate, which is recognized by the EU Commission to demonstrate compliance with the European sustainability requirements for biofuels.

The produced bio-oil is sold to FrieslandCampina that uses the bio-oil as boiler fuel in their milk powder production site in Borcula as a substitute for natural gas in order to reduce GHG emissions.

The production capacity of the EMPYRO plant will gradually increase to its maximum of over 20 million liters of bio-oil annually.

3.3.2 Biochemical Conversion

Theoretically, the source of sugar is irrelevant for the fermentation process (whether derived from sugar cane, sugar beet, or lignocellulose), provided the sugar solution contains a suitable composition of sugars and does not contain inhibitors. However, specific microorganisms have specific requirements; they cannot use all sugars, are differently robust and produce distinguished products. Therefore, there are a number of possible combinations of feedstock, pretreatment options, and microorganisms, sometimes even to produce the same product.

3.3.2.1 Bioethanol Production

The use of lignocellulose-derived sugars is driven by the increasing demand for 2G bioethanol. Producing suitable sugar solutions from lignocellulosic feedstock, especially for the production of biochemicals, is still a challenge, although significant progress has been made in recent years. A number of demonstration plants for the production of 2G bioethanol have been built in order to investigate scale-up effects. The majority of these plants, mentioned in Table 3.2, use either hydrothermal or steam-assisted pretreatments.

In addition to the initiatives listed in Table 3.2, there are several other industrial initiatives focused on the production of 2G ethanol. In 2012, Beta Renewables inaugurated the world's first commercial biorefinery plant in northern Italy, producing 40,000 t/year bioethanol from straw and *Arundo donax*. In Brasil, a Beta Renewables plant producing 65,000 t/year bioethanol was built in cooperation with GranBio. The construction of a similar biorefinery by Beta Renewables in North Carolina, USA, is ongoing. Novozymes and Beta Renewables also announced their plans to construct a cellulosic ethanol facility in India. A joint venture between Anhui Guozhen Co and Beta Renewables plans to convert 970,000–1,300,000 t/year of agricultural residues into cellulosic ethanol, glycols, and byproducts such as lignin in Fuyang City (Anhui Province, PRC).

3.3.3 Biorefineries: Starch/Sugar-Based

Biobased chemicals and biofuels must compete on cost and performance with petrochemicals and petroleum fuels. At present, most chemical-producing biorefineries use starch or sugar as the major feedstock.

BASF, Novozymes, and Cargill jointly developed *3-hydroxypropionic-acid* a raw material for the production of biobased acrylic acid, in pilot scale.[4] On the other hand, OPX Biotechnologies with its partner Dow Chemicals is aiming to commercialize bio acrylic acid by 2017. Their process is based on the direct conversion of dextrose (corn) or sucrose

[4] http://www.auri.org/assets/2014/09/AIC185.biobased1.pdf

Table 3.2 Selected second-generation ethanol operational in 2015 exceeding 1000 metric t/year of product[a]

Plant owner	Location	Technology	Feedstock	Output capacity
Clariant (ex Süd Chemie)	Straubing, Germany	Hydrothermal pretreatment at mild process conditions (Sunliquid process)	Cereal straw, corn stover, sugar cane bagasse	1000 t/year
Abengoa Bioenergy	Hugoton, Kansas, USA	Hydrothermal pretreatment (dilute acid)	Stover, wheat straw, switchgrass	74,000 t/year
Inbicon (Dong Energy)	Kalundborg, Denmark	Hydrothermal pretreatment		4300 t/year
Beta Renewables (JV Chemtex (M&G), TPG, Novozymes)	Crescentino, Italy	Two steps, steam assisted pretreatment	Straws, giant reed	60,000 t/year
Iogen	Canada	Hydrothermal pretreatment		1600 t/year
Ineos-Bio	Indian River County, Florida, USA	Gasification and fermentation	Vegetative and yard waste	24,000 t/year
Dupont Danisco	Nevada, Iowa, USA	Biochemical	Corn stover	90,000 t/year
Poet-DSM	Emmetsburg, Iowa, USA	Biochemical	Corn stover	60,000 t/year
Abengoa	Spain	Hydrothermal pretreatment	Wheat straw?	4000 t/year
GeoSynFuels[b]	USA	Biochemical	Various	4500 t/year
Quad-County	USA	Biochemical	Corn kernel fiber	6000 t/year
Gevo	USA	Biochemical		54,000 t/year
Borregard	NO	Biochemical	Residues	400,000 t/year
Chempolis	FI	Biochemical	Residues	5000 t/year
Longlive Bio-tech	CN	Biochemical	Residues	50,000 t/year
Shandong Zesheng	CN	Biochemical	Residues	3000 t/year
Cane Technology	BR	Biochemical	Bagasse	40,000 t/year
RaizenEnergia	BR	Biochemical	Bagasse	32,000 t/year

[a]Most information taken from www.biofuelstp.eu/spm5/pres/monot.pdf and http://demoplants.bioenergy2020.eu/, accessed July 2015.
[b]Demoplant brought from Blue Sugars Corporation in 2014.

(cane) through genetically engineered E-Coli1/2 via OPXBIO EDGE Technology Platform. The global acrylic acid market is forecast to reach US$18.8 billion by 2020 from US$11 billion in 2013, corresponding to a global consumption of acrylic acid of 8.2 million t/year.[5] OPXBio said it was already able to produce 38 cents/lb bioacrylic acid using sucrose feedstock at 8 cents/lb. The company plans to have a demonstration plant with a capacity of 600,000 lb/year in 2013.[6] A commercial plant with a capacity of 100 million lb/year is expected by 2015.

Succinic acid has received increasing attention for the production of new polyesters with good mechanical properties combined with full biodegradability. The main actors are probably Succinity GmbH (a Corbion Purac and BASF joint venture) with a commercial production facility in Montmeló, Spain, and Reverdia (a joint venture between Roquette and DMS) with a commercial facility in CassanoSpinola, Italy. Both plants have a production capacity of about 10,000 t/year. In Canada, BioAmber, in collaboration with Mitsui & Co., is constructing a biobased succinic acid production plant at a bioindustrial park owned by Lanxess in Sarnia, Ontario. The plant will initially have a production capacity of 17,000 t/year. The capacity will be increased by a further 35,000 t/year and at the same site it is scheduled to produce 23,000 t/year of 1,4-butanediol.[7] The estimated cost for the construction of the plant is US$110 million. For the production of succinic acid BioAmber wants to use a variety of feedstock, for example, corn, wheat, cassava, rice, sugarcane, sugar beet, and forest waste. The feedstock will initially be transported by trucks to a wet mill facility, where it will be processed into glucose (dextrose) sirup and other valuable coproducts. The glucose will then be transported to the Sarnia plant, where it will be fermented to raw biobased succinic acid and later purified to crystalline biobased succinic acid before being shipped. Costs of succinic acid production strongly depend on the plant size. Some investigations indicate a cost of US$ 2.20 (€ 1.96) per kg at the 5000 t/year. Liu (2000)estimated that the price can be lowered to US$ 0.55 (€ 0.50) per kg at a production capacity of 75,000 t/year. At present, succinic acid is mostly produced by the chemical process from

[5] http://www.researchandmarkets.com/reports/2911099/global-acrylic-acid-market-derivatives-types

[6] http://www.icis.com/blogs/green-chemicals/2012/01/first-2012-post-opxbio-update/

[7] http://investor.bio-amber.com/2011-11-08-BioAmber-and-Mitsui-amp-Co-to-Build-and-Operate-Plants-Producing-Succinic-Acid-and-BDO

Table 3.3 Estimated prices of biobased products and market volumes as recently assessed in a final report for the European Commission (E4tech et al., 2015)

Product	Price ($ per t)	Volume (1000 t/year)	% of total market
Bioethanol	815	71,310	93
BDO	>3000	3.0	0.1
n-butanol	1890	590	20
Succinic acid	2940	38	49
Xylitol	3900	160	100
Farnesene	5581	12	100

n-butane through maleic anhydride. The incidence of the raw material cost can be estimated in US$ 1.027 per kg. On the other hand, glucose is sold at a price of about $0.39 per kg, and assuming the succinic acid yield of 91% (w/w) on glucose, the raw material cost in the bioprocess is then $0.428 per kg (Song and Lee, 2006). Thus fermentative production of succinic acid from renewable resources can compete with the chemical process.

NatureWorks produces *polylactic acid* (PLA) at a plant with a designed capacity of 140,000 t/year in 2003; its current capacity is 150,000 t/year. Total investment costs have been US$ 300 million in hardware and US$450 million in R&D, process development, and technical support.[8] The current price of PLA is higher than the average cost of polyolefin (0.4–0.7 € per kg).[9] On an industrial scale, producers are seeking a target manufacturing cost of lactic acid monomer to less than US$ 0.8 per kg because the SP of PLA should decrease roughly by half from its present price of US$ 2.2 per kg.[10] This example shows that biobased materials are perceived as greener and therefore bought despite a higher price.

The European Commission Directorate-General Energy has released a report titled "From the Sugar Platform to Biofuels and Biochemical," where product prices and market volumes are depicted; some of these are shown in Table 3.3.

Selected chemicals produced at a large scale and briefly described above are shown in Table 3.4.

[8] http://www.icis.com/resources/news/2009/08/03/9235368/obama-s-green-message-boosts-biopolymer-profile/, accessed July 2015
[9] http://plasticker.de/preise/preise_monat_multi_en.php, accessed September 2015
[10] http://www.biopreferred.gov/files/WhyBiobased.pdf

Table 3.4 Selected biobased chemicals produced at a large scale

Biobased products	Plant capacity (t/year)		Company	Investment costs (range)	Preferred feedstock	Competing technology
	Minimum	Maximum				
Acrylic acid	272,000 t/year (30,000 L)	450,000 t/year (≈4 million liter)	OPXBIO-Dow Chemicals		Destrose (corn), saccarose (cane)	Petroleum-based acrylic acid
1,3-propanediol (PDO)		75,000 t/year	Dupont		Dextrose (corn)	Petroleum-based product
Iso-butanol			Gevo68 million liter		Corn, wheat, sorghum, barley and sugar cane	
Succinic acid	10,000 t/year	35,000 t/year	BIOAMBER[a]	US$ 110 million for 17,000 t/year	Starch based feedstock corn, wheat, cassava, rice, sugarcane, sugar beets and forest waste	
Succinic acid	10,000 t/year	35,000 t/year	Myriant	US$ 80 million for 14,000 t/year	Sorghum grain and grits; potentially sugars from corn stover and wood chips	
LCA/PLA	100,000 t/year	140,000 t/year PLA, (180,000 t/year LA)	Cargil-Dow	US$ 300 million+ US$ 450 million in R&D, process development	Mainly maize, cassava, rice, but also sugar beet in Europe. Future feedstock corn stover	Substitute for some polyolefins, or acrylates (Batelle) as feedstock for the chemical industry

[a]Under construction.

3.4 SUMMARY AND OUTLOOK

Results presented in this review are based on a diverse array of literature studies that include numerous, and sometimes differing, assumptions. The data quality of the literature sources used in this study is not homogeneous, and that causes an inherent uncertainty about the actual performance of the different pretreatment and conversion technologies when implemented at a large scale.

Fossil-based fuels and materials have a higher dependency on oil prices than their bio-counterparts. Most recently oil prices have been rather volatile, for example, in 2009 the average oil price was at US\$ 60 per barrel, in 2011 at US\$ 105 per barrel, and in January 2015 below US\$ 50 per barrel. Hence, it is extremely difficult to forecast at which oil price biobased products become cost-competitive incomparison with their fossil counterparts. As a rule of thumb cost-competitiveness of biobased products increases with increasing oil price and decreasing feedstock costs. Currently, the production of bioenergy and biofuels is subsidized in many countries (Kraxner et al., 2013). However, the utilization of biomass in biorefineries seems to provide higher value creation in the long term and is more sustainable than the production of bioenergy and/or biofuels alone. There are various strategies but there are no distinct policy drivers for the utilization of biobased chemicals, in direct contrast to the biofuels industry where various national regulations are driving rapid growth (Hermann et al., 2011). However, rapid growth in lignocellulosic feedstock utilization will require changes in the supply chain infrastructure and may also need more efficient socioeconomic and policy frameworks (Richard, 2010), for example, an expanded policy that includes biobased products provides added flexibility without compromising GHG targets (Posen et al., 2014).

Currently, no biomass pretreatment technology able to convert lignocellulosic biomass into sugars for fermentation is cost-competitive with conventional sugar. There is no single preferred pretreatment method, with mild acid hydrolysis and steam explosion, dilute and concentrated acid, and mild alkaline processes all planned to be used in biofuel plants in the near future.

Fermentation of low-cost sugar is economically the most attractive option for the production of biofuels and other biorefinery products, particularly PLA and succinic acid (Posada et al., 2013; Gerssen-Gondelach et al., 2014). The integration of different pretreatment and conversion technologies in biorefineries can maximize the use of all biomass components and improve both the economic and the ecological efficiency of the whole value chain. However, eco-efficient biobased value chains call for different ways of thinking about agriculture, energy infrastructure, processing industry, and rural economic development. A key challenge for developing and growing a commercially secure bioeconomy is a reliable and consistent feedstock supply.

In the longterm, thermochemical and biochemical conversion of lignocellulosic biomass are promising technologies for the production of biofuels and biobased chemicals. Realizing the long-term opportunities of lignocellulosic biomass conversion depends both on technological learning and fossil fuel prices (Saygin et al., 2014).Some chemical companies may also be keen to decouple their base and intermediate materials from fossil-based sources.

At present, the bioeconomy is a political concept and efforts from all actors are required for its realization. The pace of large-scale implementation of biorefineries depends not just on the environmental and economic performance of pretreatment and conversion technologies, but also on governmental policy framework and societal preferences. It is unlikely that a single solution or policy will determine the extent to which biorefineries are incorporated into a global biobased economy. Rather, it is likely that a robust portfolio of solutions, appropriate for site-specific conditions (eg, feedstock availability, supply chain infrastructure, rural development strategies, etc.), will be needed to sustainably address the future material and energy demand (Wilcox, 2014).

REFERENCES

Aden, A., 2008. Biochemical Production of Ethanol From Corn Stover: 2007 State of Technology Model. NREL, Golden, USA.

Aden, A., 2009. Biochemical Production of Ethanol From Corn Stover: 2008 State of Technology Model. NREL, Golden, USA.

Anex, R.P., Aden, A., Kazi, F.K., Fortman, J., Swanson, R.M., Wright, M.M., et al., 2010. Techno-economic comparison of biomass-to-transportation fuels via pyrolysis, gasification, and biochemical pathways. Fuel 89 (Suppl. 1), S29–S35.

Argo, A.M., Tan, E.C.D., Inman, D., Langholtz, M.H., Eaton, L.M., Jacobson, J.J., et al., 2013. Investigation of biochemical biorefinery sizing and environmental sustainability impacts for conventional bale system and advanced uniform biomass logistics designs. Biofuel. Bioprod. Bior. 7 (3), 282–302.

Benali, M., Périn-Levasseur, Z., Savulescu, L., Kouisni, L., Jemaa, N., Kudra, T., et al., 2014. Implementation of lignin-based biorefinery into a Canadian softwood kraft pulp mill: optimal resources integration and economic viability assessment. Biomass Bioenerg. 67, 473–482.

Bioliq, 2014. Available from: <http://www.bioliq.de/english/55.php> (accessed 16.12.14.).

Bridgwater, A.V., 2009. Technical and Economic Assessment of Thermal Processes for Biofuels. NNFCC, York.

Brown, T.R., 2015. A critical analysis of thermochemical cellulosic biorefinery capital cost estimates. Biofuel. Bioprod. Bior. 9 (4), 412–421.

Cavalett, O., Junqueira, T., Dias, M.S., Jesus, C.F., Mantelatto, P., Cunha, M., et al., 2012. Environmental and economic assessment of sugarcane first generation biorefineries in Brazil. Clean Technol. Environ. Policy 14 (3), 399–410.

Cheali, P., Posada, J.A., Gernaey, K.V., Sin, G., 2015. Upgrading of lignocellulosic biorefinery to value-added chemicals: sustainability and economics of bioethanol-derivatives. Biomass Bioenerg 75, 282–300.

Choudhary, V., Sandler, S.I., Vlachos, D.G., 2012. Conversion of Xylose to Furfural Using Lewis and Brønsted Acid Catalysts in Aqueous Media. ACS Catal. 2 (9), 2022–2028.

Dahmen, N., Dinjus, E., Kolb, T., Arnold, U., Leibold, H., Stahl, R., 2012. State of the art of the bioliq (R) process for synthetic biofuels production. Environ. Progress Sustaina. Energ. 31 (2), 176–181.

Daugaard, T., Mutti, L.A., Wright, M.M., Brown, R.C., Componation, P., 2015. Learning rates and their impacts on the optimal capacities and production costs of biorefineries. Biofuel. Bioprod. Bior. 9 (1), 82–94.

Davis, R., Tao, L., Tan, E.C.D., Biddy, M.J., Beckham, G.T., Scarlata, C., et al., 2013. Process Design and Economics for the Conversion of Lignocellulosic Biomass to Hydrocarbons: Dilute-Acid and Enzymatic Deconstruction of Biomass to Sugars and Biological Conversion of Sugars to Hydrocarbons. National Renewable Energy Laboratory, Idaho National Laboratory, Harris Group Inc, Golden, CO, USA.

E4tech, RE-CORD and WUR, 2015. From the Sugar Platform to Biofuels and Biochemicals. Final report for the European Commission.

EMPYRO, 2014. Available from: <http://www.empyroproject.eu> (accessed 17.12.14.).

ENSYN, 2014. Available from: <http://www.ensyn.com> (accessed 16.12.14.).

García, A., Alriols, M.G., Llano-Ponte, R., Labidi, J., 2011. Energy and economic assessment of soda and organosolv biorefinery processes. Biomass Bioenerg. 35 (1), 516–525.

Gerssen-Gondelach, S.J., Saygin, D., Wicke, B., Patel, M.K., Faaij, A.P.C., 2014. Competing uses of biomass: assessment and comparison of the performance of bio-based heat, power, fuels and materials. Renew. Sustain. Energ. Rev. 40 (0), 964–998.

Hamelinck, C.N., Suurs, R.A.A., Faaij, A.P.C., 2005. International bioenergy transport costs and energy balance. Biomass. Bioenerg. 29 (2), 114–134.

Hermann, B., Carus, M., Patel, M., Blok, K., 2011. Current policies affecting the market penetration of biomaterials*. Biofuel. Bioprod. Bior. 5 (6), 708–719.

Höltinger, S., Schmidt, J., Schönhart, M., Schmid, E., 2014. A spatially explicit techno-economic assessment of green biorefinery concepts. Biofuel. Bioprod. Bior. 8 (3), 325–341.

Hytönen, P.R.S., 2009. Integrating bioethanol production into an integrated kraft pulp and paper mill: techno-economic assessment. Pulp Pap. Canada. 1, 25–32.

Kazi, F.K., Fortman, J.A., Anex, R.P., Hsu, D.D., Aden, A., Dutta, A., et al., 2010. Techno-economic comparison of process technologies for biochemical ethanol production from corn stover. Fuel 89 (Suppl. 1), S20–S28.

Kraxner, F., Nordström, E.-M., Havlík, P., Gusti, M., Mosnier, A., Frank, S., et al., 2013. Global bioenergy scenarios – Future forest development, land-use implications, and trade-offs. Biomass Bioenerg. 57 (0), 86–96.

Kumar, D., Murthy, G.S., 2011. Pretreatments and enzymatic hydrolysis of grass straws for ethanol production in the pacific northwest U.S. Biol. Eng. 3 (2), 97–110.

Lang, H.J., 1947. Engineering approach to preliminary cost estimates. Chem. Eng. 57 (9), 130–133.

Lang, H.J., 1948. Simplified approach in preliminary cost estimates. Chem. Eng. 55 (6), 112–113.

Liu, Y., 2000. New Biodegradable Polymers From Renewable Resources. Royal Institute of Technology, Stockholm, Sweden, PhD.

Manitoba, 2014. Available from: <http://www.hydro.mb.ca/your_business/bioenergy_optimization/demonstrations.shtml> (accessed 16.12.14.).

Meier, D., van de Beld, B., Bridgwater, A.V., Elliott, D.C., Oasmaa, A., Preto, F., 2013. State-of-the-art of fast pyrolysis in IEA bioenergy member countries. Renew. Sustain. Energ. Rev. 20, 619–641.

Muth, D.J., Langholtz, M.H., Tan, E.C.D., Jacobson, J.J., Schwab, A., Wu, M.M., et al., 2014. Investigation of thermochemical biorefinery sizing and environmental sustainability impacts for conventional supply system and distributed pre-processing supply system designs. Biofuel. Bioprod. Bior. 8 (4), 545–567.

O'Keeffe, S., Schulte, R.P.O., Sanders, J.P.M., Struik, P.C., 2011. I. Technical assessment for first generation green biorefinery (GBR) using mass and energy balances: scenarios for an Irish GBR blueprint. Biomass Bioenerg. 35 (11), 4712–4723.

O'Keeffe, S., Schulte, R.P.O., Sanders, J.P.M., Struik, P.C., 2012. II. Economic assessment for first generation green biorefinery (GBR): scenarios for an Irish GBR blueprint. Biomass Bioenerg. 41 (0), 1–13.

Peters, M.S., Timmerhaus, K., West, R., 2002. Plant Design and Economics for Chemical Engineers. McGraw-Hill, Columbus, OH.

Pinho, A. d R., de Almeida, M., Mendes, F.L., Ximenes, V.L., Casavechia, L.C., 2015. Co-processing raw bio-oil and gasoil in an FCC Unit. Fuel Process. Technol. 131, 159–166.

Posada, J.A., Patel, A.D., Roes, A., Blok, K., Faaij, A.P.C., Patel, M.K., 2013. Potential of bioethanol as a chemical building block for biorefineries: preliminary sustainability assessment of 12 bioethanol-based products. Bioresour. Technol. 135 (0), 490–499.

Posen, I.D., Griffin, W.M., Matthews, H.S., Azevedo, I.L., 2014. Changing the renewable fuel standard to a renewable material standard: bioethylene case study. Environ. Sci. Technol. 49 (1), 93–102.

Reyes Valle, C., Villanueva Perales, A.L., Vidal-Barrero, F., Ollero, P., 2015. Integrated economic and life cycle assessment of thermochemical production of bioethanol to reduce production cost by exploiting excess of greenhouse gas savings. Appl. Energ. 148, 466–475.

Richard, T.L., 2010. Challenges in Scaling Up Biofuels Infrastructure. Sci. 329 (5993), 793–796.

Ringer, M., Putsche, V., Scahill, J., 2006. Large-Scale Pyrolysis Oil Production: A Technology Assessment and Economic Analysis. NREL, Golden, CO.

Saygin, D., Gielen, D.J., Draeck, M., Worrell, E., Patel, M.K., 2014. Assessment of the technical and economic potentials of biomass use for the production of steam, chemicals and polymers. Renew. Sustain. Energ. Rev. 40, 1153–1167.

Sievers, D.A., Tao, L., Schell, D.J., 2014. Performance and techno-economic assessment of several solid–liquid separation technologies for processing dilute-acid pretreated corn stover. Bioresour. Technol. 167 (0), 291–296.

Song, H., Lee, S.Y., 2006. Production of succinic acid by bacterial fermentation. Enzyme Microbial Technol. 39 (3), 352–361.

Stanil, E., Patel, M., 2011. Recommended Methodology and Tool for Cost Estimates at Micro Level for New Technologies. Utrecht, Netherlands.

Swanson, R.M., Platon, A., Satrio, J.A., Brown, R.C., 2010. Techno-economic analysis of biomass-to-liquids production based on gasification. Fuel 89 (Supp. 1), S11–S19.

Tao, L., Tan, E.C.D., McCormick, R., Zhang, M., Aden, A., He, X., et al., 2014. Techno-economic analysis and life-cycle assessment of cellulosic isobutanol and comparison with cellulosic ethanol and n-butanol. Biofuel. Bioprod. Bior. 8 (1), 30–48.

Towler, G., Sinnott, R.K., 2007. Chemical Engineering Design: Principles, Practice and Economics of Plant and Process Design, Butterworth-Heinemann.

Turley, D., Evans, G., Nattrass, L., 2013. Use of sustainably-sourced residue and waste streams for advanced biofuel production in the European Union: rural economic impacts and potential for job creation. NNFCC, York, UK.

Wilcox, J., 2014. Grand challendges in advanced fossil fuel technologies. Front. Energ. Res. 2.

Wright, M.M., Daugaard, D.E., Satrio, J.A., Brown, R.C., 2010. Techno-economic analysis of biomass fast pyrolysis to transportation fuels. Fuel 89 (Suppl. 1), S2–S10.

Zinoviev, S., Müller-Langer, F., Das, P., Bertero, N., Fornasiero, P., Kaltschmitt, M., et al., 2010. Next-generation biofuels: survey of emerging technologies and sustainability issues. ChemSusChem 3 (10), 1106–1133.

CHAPTER 4

Sustainability Considerations for the Future Bioeconomy

R. Diaz-Chavez[1], H. Stichnothe[2] and K. Johnson[3]
[1]Centre for Environmental Policy, Imperial College, London, United Kingdom
[2]Thünen Institute of Agricultural Technology, Braunschweig, Germany
[3]US Department of Energy, Bioenergy Technologies Office, Golden, CO, United States

Contents

Abstract

It is critical to ensure the sustainability of biomass when used for energy, chemicals, and/or materials in the future bioeconomy. This does not only apply to the feedstock, a common focus within traditional bioenergy assessments; it also needs to consider the wider value chain, that is, from feedstock production through end use, including a range of coproducts, to end-of-life. The scope of such an assessment can vary but may be most practical at the "biorefinery" scale. Experience gained from first-generation biofuels offers lessons about sustainability challenges and prospects for the future bioeconomy. However, sustainability assessments of bioproducts require unique considerations, some of which are not necessarily addressed in the assessments of biofuels. We find that sustainability assessments are not "one-size-fits-all" and should engage stakeholders in determining clear goals and objectives for the assessment, consider the specific context, and maintain transparency in approach and assumptions. Sustainability is also not a steady state or fixed target. Sustainability assessments are most useful when they help decisionmakers and technology developers make continuous improvements across social, environmental, and economic dimensions. In addition to the traditional three-pillar approach, good governance is of equal importance and has to be implemented in sustainability assessment frameworks. As such, methodologies must continuously evolve to accommodate the increasingly diverse range of biomass-derived products within the future bioeconomy.

Developing the Global Bioeconomy.
DOI: http://dx.doi.org/10.1016/B978-0-12-805165-8.00004-5

© 2016 Elsevier Inc.
All rights reserved.

4.1 INTRODUCTION

Biorefineries transform biomass resources into a variety of products that can be used or recycled as a material or energy. Biorefineries can affect social, environmental, and economic wellbeing, three dimensions often considered in sustainability assessments. Within the bioeconomy, biomass will be used for the sustainable production of food and feed, as well as chemicals, materials, and energy (power, heating/cooling, and/or transport). The total value of the biomass feedstock can be maximized through a so-called "biorefinery approach," which integrates conversion processes and equipment to coproduce multiple products from different biomass components and intermediates.

Biorefineries can be subdivided into energy-driven biorefineries and product-driven biorefineries. Energy-driven (or biofuel-driven) biorefineries produce mainly huge volumes of relatively low-value energy (or fuels) out of biomass. The full-value chain infrastructure exists; however, at current fuel prices their profitability is still questionable, requiring significant financial governmental support or a regulated market to guarantee large-scale market deployment. Product-driven (ie, chemicals, materials) biorefineries typically produce smaller amounts of relatively higher value-added biobased products out of biomass; primary (agro) and secondary (process) residues are used to produce energy (power/heat) for internal or external use. Currently, only limited product-driven biorefineries are in operation. Biorefineries, if appropriately designed and operated, contribute to sustainable innovation and may foster development in rural areas.

Sustainability assessment is context-specific and subject to change across temporal and spatial scales in response to changing societal needs, economics, and environmental conditions. There is no "one-size-fits-all" sustainability assessment method. The type of assessment is determined by the purpose and objective of the sustainability assessment. Sustainability assessment can serve multiple purposes, but each has certain requirements regarding relevant aspects, data quality, methodological choices, and constraints such as data availability, resource restrictions, and confidentiality issues. Stakeholders should play a central role in defining goals and selecting indicators that are appropriate for a particular assessment. If the purpose of the assessment is the comparison of two options (technologies or products), all relevant indicators for each option should be considered. The indicators for the comparison should be derived using a consistent approach, consistent system boundaries, and relevant data (ISO, 2015).

Sustainability is not a steady state or fixed target. Assessing sustainability involves comparing the relative merits of different options, and achieving it allows for continued adjustment in response to changing conditions, knowledge, and priorities. The concept and assessment has evolved from an environmental position to an integrated one including environmental, economic, social, and policy/institutions or good governance (Diaz-Chavez, 2015). Sustainability assessment is a tool that uses different methods to emphasize synergistic, adverse, as well as short- and long-term effects of different alternatives (OECD, 2008). Several of these methods including traditional environmental management tools (eg, environmental and social impact assessment, strategic environmental assessment) are based on multicriteria methods and the use of indicators for measuring the currently more modern approach of four dimensions of sustainability (GBEP, 2010; Diaz-Chavez, 2014, 2015). Sustainability assessments need to be context-specific because sustainability issues are related to the people and the area that are affected by the proposal. This is also why stakeholder participation is important. The sustainability assessment process is a learning process that should have adequate resources for constant improvements (O'Connell et al., 2013).

Fig. 4.1 shows selected aspects of sustainability issues of bioproducts that should be taken into account.

Despite the plethora of activities related to biomass and bioenergy sustainability assessments, from methodology development to standardization and certification (Buchholz et al., 2015; Keller et al., 2015; Patel et al., 2015; Pinazo et al., 2015; Fritsche and Iriarte, 2014; Buchholz et al., 2007; Martins et al., 2007; Sikdar, 2007), there is no consensus yet for a harmonized assessment framework. According to a report from the Organisation for Economic Cooperation and Development (OECD, 2010), a common framework may help governments and industry to identify, evaluate performance, and support the development of bioproducts which are likely to be more sustainable than their fossil counterparts.

Such a framework could focus on applying minimum criteria or specific methodologies. As much as possible, assessments should be carried out on a lifecycle basis, starting from biomass feedstock and extending to (and beyond) the end-of-life of the products derived from its biomass feedstocks. Consistency and transparency are key requirements for communicating sustainability results to stakeholders and the public. The target audience using sustainability assessment results includes policymakers,

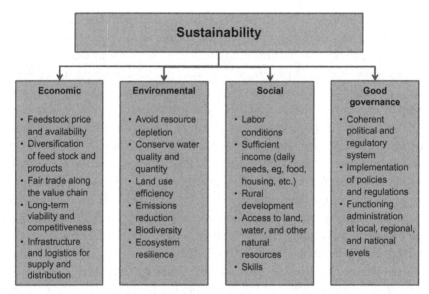

Figure 4.1 Key issues in four areas of sustainability.

business people, and other stakeholders in all stages of the supply chain from land managers or waste suppliers to those involved in logistics, conversion facilities, and end-users.

This chapter presents an overview of issues related to sustainability assessments that are applicable to the future bioeconomy and integrated biorefineries. It is not exhaustive but presents advancements and lessons learned to date.

4.2 OVERVIEW OF METHODOLOGIES AND SUSTAINABILITY ASSESSMENT FRAMEWORKS

As sustainability assessment evolved, so also have the methodologies and frameworks applied to project planning and implementation within the bioenergy sector. Different methodologies have been used to assess projects in bioenergy with a particular emphasis on feedstock production (Keam and McCormick, 2008; Rosillo-Calle et al., 2015), the supply chain (eg, Batidzirai, 2013), and the technology used or the process and pathways employed (Dewulf and Van Langenhove, 2006). Many of the methods and frameworks have focused on bioenergy production rather than bioproducts or integrated biorefineries that produce multiple products.

Many sustainability assessments use multicriteria analysis and indicators to monitor the effects of bioenergy feedstocks, technologies, and industrial developments at local, regional, or national levels. For example, the Global Bioenergy Partnership has developed environmental, economic, and social sustainability indicators that can be used by countries to inform development and monitor effects of national-level bioenergy policies and programs (GBEP, 2011). Reviews of sustainability methodologies and the indicators proposed have been published (eg, by Cherubini et al., 2009; Sacramento-Rivero, 2012; Diaz-Chavez, 2014, 2015). Several initiatives have supported sustainability assessments of biorefineries using indicators, including for instance Jungmeier (2014) and BIOCORE (2014).

Criteria and indicators are also widely used in sustainability frameworks and standards. Some of the indicators were also linked to policies and programs. For sustainability assessments to be useful, socioeconomic and environmental indicators need to target stakeholder needs (Dale et al., 2015). Obtaining sufficient evidence to show quantifiable relationships among causes and effects is a key challenge affecting the selection of indicators (Dale et al., 2013). Some effects may differ not only in magnitude but in direction, depending on how, where and when measurements are made. Some indicators can be directly measured and attributable to a supply chain (eg, employment, profitability, public reporting), while others may require considerable research to discern and allocate relative values.

Different policies and mandatory targets have influenced the production of biofuels, but few legal systems incorporate criteria or indicators that regulate sustainability. The EU Renewable Energy Directive (RED) (2009/28/EC) introduced basic environmental sustainability criteria for biofuel production including greenhouse gas (GHG) emissions, protection of land with high biodiversity value, or high carbon stock plus agroenvironmental practices. Since its adoption, several standards that cover these criteria have been approved by the EU (17 until 2015; EC, 2015).

Voluntary standards for certification of biomass products have been developed and are commonly used in the USA, for example, the Sustainable Forestry Initiative (SFI); Forest Stewardship Council (FSC); and Council on Sustainable Biomass Production (CSBP). Several benchmarking exercises have been conducted to review the similarities and differences of these systems (eg, van Dam et al., 2010).

The International Organization for Standardization (ISO) has also developed a standard on Sustainability criteria for bioenergy ISO 13065 (2015). The European Committee for Standardization (ECN) in Europe is developing a standard for biobased products (CEN/TC411) and has set up five working groups dealing with relevant issues: WG1 Terminology; WG2 Bio-based solvents; WG3 Biobased content, biological origin, measurement methods; WG4 LCA and sustainability of biobased products; and WG5 Certification and communication (Biobased Economy, 2015). While these diverse mandatory and voluntary approaches have helped establish consensus about the importance of addressing sustainability associated with energy production and use, there is still little agreement on practical steps for decisionmakers to evaluate the relative sustainability of different energy options.

One approach to assessing biobased projects involves lifecycle analysis that subdivides a process into several stages, for example, raw material provision, conversion, consumption, and end-of-life (Fig. 4.2). When using a lifecycle perspective for assessing the environmental dimensions of biomass- or petroleum-derived products, many process steps, such as purification, product formulation, storage, and waste water treatment, are similar, while others are distinct (such as establishing the fuel source, obtaining materials, distributing materials to refineries, and converting materials into fuel) (Parish et al., 2013). The same may apply to assessments of economic and social aspects, although the system boundaries may vary. A fourth dimension shown in Fig. 4.1 indicates the importance of contextual conditions related to governance including policy, regulations, enforcement, and institutional capacity assessment which looks into the area of policy support, governance, and institutions.

Keller et al. (2015) provided a review of methodologies used specifically on the integrated sustainability assessment for biorefineries based on lifecycle assessment. The methodology of integrated lifecycle sustainability assessment (ILCSA) utilizes existing methodologies and added features for exante assessments. This approach allowed flexibility for focusing on those sustainability aspects relevant in the decisionmaking process using the best available

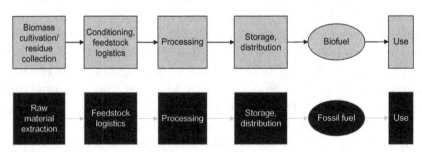

Figure 4.2 Value chain of biofuels and fossil fuels.

methodology for assessing each aspect within the overarching assessment. Environmental and social-economic dimensions have also been analyzed individually in combination with other environmental management tools (Diaz-Chavez, 2014). The IEA Task 42 on Biorefineries produced another framework based on factsheets to assess biorefineries. The system uses a description of the key attributes of a biorefinery; based on a technical description and a classification scheme, the mass and energy balance is calculated for the most reasonable production capacity. Then three sustainability dimensions (as depicted in Fig. 4.2, except for "good governance") are assessed and documented in a compact form in the "Biorefinery Fact Sheet." This framework has been applied to multiple types of biorefineries including different feedstock and chemical platforms (Jungmeier, 2014).

Sustainability assessments of biomass-derived chemicals and materials present new challenges relative to more straightforward analyses comparing biofuels versus fossil fuels. A material's service life is more complex, and, in addition, the consumer behavior has a significantly higher impact on a product's overall sustainability. While biofuels have a dedicated combustion application, chemicals/materials (either standalone or embedded in more complex products) can have different service life times and end-of-life destinations. The former depends on consumer behavior while the latter depends on both consumer behavior and the existing waste management infrastructure, which influence whether materials are reused, recycled, downcycled, landfilled, or discarded. Fuels just have a single life, as illustrated in Fig. 4.2.

The main difference when comparing fuels is the feedstock choice. Regional and temporal variations in crop yield and residue availability and prices complicate the sustainability assessment, and there is a general lack of regional-specific data that can be used to evaluate biomass availability in a specific region/country.

4.3 LESSONS LEARNED FROM FIRST-GENERATION BIOFUELS AND BIOENERGY CROPS

Since 1990, the carbon and fossil energy footprints of corn production have significantly decreased due to higher corn yields and improved agricultural practices (Babcock, 2015; Chum et al., 2015). Simultaneously, corn ethanol mills have evolved leading to a substantial reduction in production costs, fresh water demand, fossil energy consumption, and GHG emissions associated with ethanol production (Chum et al., 2015). These gains were due to a combination of industry learning, adoption of more energy-efficient technologies, streamlined processing configurations, increased mill size, and replacing coal with lower-carbon process energy. As both corn and ethanol yields have increased, the land-use intensity of corn ethanol has consistently declined (Chum et al., 2013). Growth in the US corn ethanol industry also generated social and economic benefits by catalyzing investments in agricultural R&D, infrastructure, and rural communities (Horta Nogueira et al., 2015). While the environmental performance of first-generation biofuels has improved over the years, biofuels have been subjected to continuing criticism, primarily around concerns that feedstock production may compete with other land-use options, such as food production or habitat, potentially causing direct or indirect land-use change (LUC and iLUC) (Boddiger, 2007; Searchinger and Heimlich, 2015).

Since the early 2000s, there have been a number of studies applying modeling approaches to estimate LUC and iLUC of biofuel production and the associated GHG emissions (Macedo et al., 2015). Early estimates projected high negative impacts; however, these models included simplistic representations of the drivers of LUC and did not account for factors such as yield improvements, multicropping, and double-cropping, and use of underutilized pastureland (Souza et al., 2015).

Model improvements to account for these complexities, as well as comparisons to empirical land-use data, have resulted in much lower iLUC and GHG estimates in recent years (Taheripour and Tyner, 2013). Consistent with these model improvements, analyses of recent empirical land-use data suggest that increased US corn production was largely achieved with intensification on existing agricultural acres—through multiple cropping and technology improvement—rather than cultivating more land (Babcock, 2015). These examples illustrate how bioenergy markets can promote more efficient agricultural and forestry practices, and that care must be taken when applying models and interpreting results about the complex interactions between markets, land-use decisions, and environmental impacts. It is critical that modeling studies test and validate underlying assumptions with empirical data (Youngs and Somerville, 2014; Panichelli and Gnansounou, 2015; Plevin et al., 2015).

The evolution of the US corn ethanol industry offers several lessons that are relevant to the future bioeconomy. Corn ethanol expansion generated increasing concerns about negative environmental and social consequences, largely focused on LUC, impacts to food production, and the intensive agricultural practices associated with corn production. These concerns spurred a number of modeling and research efforts to quantify these impacts, and several lessons emerge from this body of knowledge: the contributions of research, development, and technological learning in improving environmental and energy performance; the importance of scientific rigor when conducting and interpreting modeled analyses on land-use changes and other impacts; and the value of investigating trends over time when evaluating sustainability (Souza et al., 2015).

Modeling and assessing are more easily carried out for biofuel-driven biorefineries as fuels have relatively short value chains. Biochemicals and biomaterials however are typically intermediate products and become part of considerably longer, more complex value chains. Consequently, assessments of product-driven biorefineries are often partial evaluations that are limited by the amount of available data. Another complicating factor is that biorefineries are highly diverse in form and still a nascent industry. As such, there are limited data available. Various examples illustrate the wide range in design configurations and product mixes (see Souza et al., 2015).

4.4 SUSTAINABILITY ASSESSMENT CHALLENGES

In the case of biomass utilization for bioenergy, including liquid biofuels, and for the bioeconomy, several challenges have been identified that draw, among others, from the lessons derived from first-generation biofuels production. Some of the key challenges are related to food security, land use change, genetically modified organisms (GMO), policy, biodiversity, and GHG balance (Mohr and Raman, 2013). Some argue that these challenges may be exacerbated in the production of second-generation biofuels (Mohr and Raman, 2013). Nevertheless, in a future bioeconomy, some of these challenges may also be overcome when using agricultural, forestry, or processing residues. Field-level studies have shown that the use of excess residues can be done sustainably (Muth et al., 2012, 2013). As these residues have no use at this point, products derived from such feedstock could substitute others (eg, derived from food crops or fossil origin) and reduce stress on land. In this section, we discuss the most prevalent potential sustainability challenges and their role in the future bioeconomy.

a. A major challenge is addressing the controversial issue of using *food biomass* for biofuel production due to the perceived impacts on food security from direct and/or indirect land use change. A recent analysis (Souza et al., 2015) counters these concerns, but results are not yet widely distributed.

While using food biomass has a (direct or indirect) impact on linked markets and, therefore, biomass prices (Heijungs et al., 2010), bioenergy from food crops can promote stable prices and thereby incentivize local production. Improved food security results from predictable and stable prices that create incentives for local investment in food production (IFPRI, 2015). Food (in)security strongly relates to household income, because welfare measurements are indicated by the fraction of marginal income spent on food (FAO, 2011). Other additional issues affect food production, security, and availability (Diaz-Chavez, 2010; Rosillo-Calle and Johnson, 2010).

Also, the discussion of how biomass production relates to food security should be accompanied with equal attention to the impact of fossil fuel development on food systems. The extraction of fossil-based feedstock, for example, lignite, tar sands, or via fracking also occupies land and contaminates water sources (Warner et al., 2013; Johnson et al., 2015; Raman et al., 2015; Sherval, 2015) which can potentially be used for irrigation and therefore may have indirect effects on food production systems. These aspects are mostly neglected in comparative sustainability assessment of fuels, which focus mostly on GHG emissions.

While estimates of future sustainable bioenergy potential vary greatly based on dynamic considerations, numerous studies demonstrate the technical potential for increasing biomass production while simultaneously meeting other goals for food security, climate change mitigation, and ecosystem health (Souza et al., 2015).

b. Concerns about the use of *GMOs* as feedstocks or in biocatalytic conversion processes also presents challenges. Perceptions and definitions of GMOs vary widely in different parts of the world. For example, in some parts of South America, GMO crops are more widely used as conventional breeds, while in parts of Europe and India, growing GMO crops is restricted by law. These differences are directly related to the general perception within the population: some consider GMOs to be critical for meeting global needs for food and other agricultural commodities, but there are also concerns that ecosystems and biodiversity could be seriously damaged. Hence, transparency is key to communicating the use of GMO crops to stakeholders and the public. Assessing the impact on ecosystem services, including biodiversity in particular, is still an unresolved problem in environmental sustainability assessment, and further research is needed in this area.

c. Allocating *social impacts and good governance* associated with bioenergy is also challenging (IEA-Bioenergy, 2015). Determining influences of a growing bioeconomy on socioeconomic indicators is particularly vexing because social conditions vary greatly and depend on many different factors (Luchner et al., 2013). Attributing social effects to particular causes is always difficult, and attributing particular effects to biorefinery products is likely to be impossible in situations where good governance is not established, laws and regulations are lacking, or when human rights are abused. Nevertheless, reviews on social issues besides job creation and working conditions have been conducted and include issues such as health and safety, competition of crops or residues and intermediate products with other uses, land uses and tenure (at the feedstock production level), and social acceptability of new products (see Diaz-Chavez, 2014; Raman et al., 2015).

d. Calculating the *GHG balance* of bioproducts with a long service life, for example, construction, is particularly challenging (Miner et al., 2014; Ter-Mikaelian et al., 2015). The key issue in the use of sustainably produced biomass for energy focuses on the timing of mitigation benefits, not whether they exist (Helin et al., 2013; Marland et al., 2013; Buchholz et al., 2014). It is difficult to estimate the lifetime of the bioproducts, that is, the respective carbon storage/sequestration time. There is also limited understanding of future feedstock options or prices of derived bioproducts. If bioproducts are used after their service life, for example, for energy recovery, a GHG assessment must make an assumption about which energy carrier is substituted, which may vary on the respective country's or region's energy mix (and related GHG emission factor), or the merit order of the energy produced (eg, in the case of biopower). For example, substituting coal would have much greater impact on the GHG calculation than substituting solar power. Despite these challenges, calculating GHG balances of long-lived bioproducts is possible as illustrated by several analyses of bioproducts obtained from lignocellulosic residues and dedicated crops (BIOCORE, 2014).

e. The *lifecycle of biomass-derived chemicals/materials* is far more complex than the lifecycle of biofuels (Fig. 4.3). While the lifecycle is similar up to the first processing level, various downstream options are possible, such as different processing technologies and use of intermediates in existing refineries, for example, to produce blended bioproducts. The intermediates can also be further processed to more sophisticated products. Bioproducts can be re-used, recycled at different stages of the

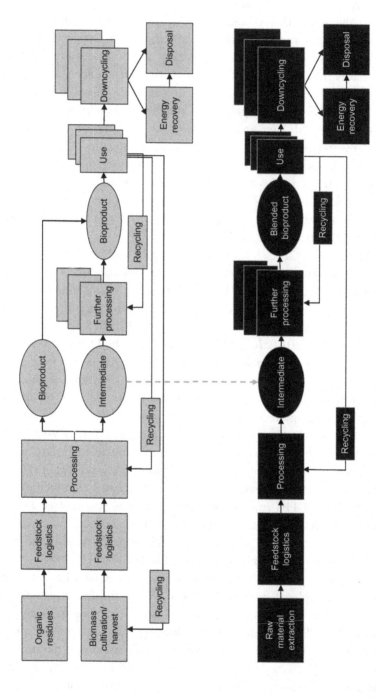

Figure 4.3 Lifecycle stages of products derived from fossil resources (in black boxes) and from biomass (in gray boxes).

value chain, downcycled (eg, from food packaging to plastic chairs), or incinerated to recover energy. They could also be disposed or released into the environment. Each option has different economic, environmental, and social consequences. Whether a bioproduct can and will be recycled depends on several factors such as governance, existing infrastructure, and consumer behavior. How efficiently bioproducts will be recycled depends on the applied technology and management practice. Other methodological challenges include the setting of system boundaries, the access to or lack of data, value judgments, and ethical issues among others (McManus et al., 2015).

f. Understanding and quantifying the benefits of a *circular economy* remains a challenge. The circular economy focuses on efficient use of finite resources and ensures that those are reused or recycled as long as possible. In the case of the bioeconomy, it promotes the integration of renewable resource production, in particular renewable carbon, and facilitates the recycling of carbon after efficient uses. In the EU, this is stated in the Action Plan for the Circular Economy (EC, 2015). The bioeconomy is a circular economy as it regenerates CO_2 and uses renewable raw materials to make greener everyday products such as food, feed, fibers, chemicals, materials, fuels, and energy. As stated in the EU Bioeconomy Strategy: "It is considered that the bioeconomy is circular by nature because carbon is sequestered from the atmosphere by plants. After uses and reuses of products made from those plants, the carbon is cycled back as soil carbon or as atmospheric carbon once again" (Bioconsortium, 2015). However, more real-life examples and additional research are needed to fully understand and quantify the benefits of a circular economy.

g. As research in bioenergy and the bioeconomy progresses, *policies* may require updates and revisions in response to new scientific information and learnings. An example is the iLUC policy in the EU. The targets set forth in the RED 2009/28/EC have been changed as the Agricultural and Fisheries Council of the European Union officially adopted the new rules to address iLUC impacts associated with biofuels. The updated 2015 directive places a 7% cap on conventional biofuels that can count towards the RED targets and allows member states to set a lower cap. It encourages the transition to advanced biofuels, includes a provision for double counting of feedstocks for advanced biofuels, and requires reporting on GHG emission savings from the use of biofuels to be carried out by the fuel supplier and the EU Commission

(Voegele, 2015). It should be noted, though, that the EU post-2020 policies on biofuels remain unclear, as the energy and climate strategy for 2030 does not specify any specific bioenergy or biofuel target. Furthermore, beyond R&D, there is no specific EU policy to foster the bioeconomy or its sustainability.

4.5 CONSIDERATIONS FOR FUTURE ASSESSMENTS IN THE BIOECONOMY SECTOR

Combating climate change, conserving biodiversity, ensuring energy security, conserving fossil resources, revitalizing rural areas, and addressing the need for economic growth without environmental and social damage are key challenges for sustainable development, as specified recently in the UN Sustainable Development Goals (UN, 2015). Using biomass as an industrial raw material, replacing fossil/conventional alternatives, is one way towards a more sustainable future. The availability of sustainable biomass as a future substitute for fossil resources is dependent on the availability of land and options for using biomass and residues produced in agriculture and forestry more efficiently. The production of biobased products can contribute to value creation and the preservation of jobs, particularly in rural areas (German Government, 2012).

Applying landscape design principles offers a means to realize the benefits of biomass utilization, bioenergy, and biomaterials while meeting other environmental, social, and economic goals if sustainability requirements are considered. Landscape design involves developing a collaborative, spatially explicit plan to manage landscapes and supply chains. A landscape design process as suggested by Dale et al. (2016), involving six steps, considers both the benefits that can be derived as well as potential negative impacts. Several principles can help avoid, reduce, or mitigate negative outcomes: conserving ecosystem and social services, recognizing the specific local and regional context of the bioenergy system, and employing adaptive management by monitoring outcomes and adjusting plans to improve performance over time (Dale et al., 2016). It is important to note several constraints and considerations for implementing landscape design, such as the challenge of applying the landscape design approach at large scales where the diversity of stakeholders and landowner objectives makes it difficult to develop common goals (Dale et al., 2016).

Several innovations can support the process of sustainably integrating feedstock production into existing landscapes. For example, on-the-ground research is demonstrating how energy crops could enhance water quality, soil carbon, or other ecosystem services while minimizing concerns about land use change (Iqbal et al., 2014; Ssegane et al., 2015). Also useful are optimization methods that use spatial data and knowledge of bioenergy attributes to present the highest-impact scenarios for established objectives (Cibin and Chaubey, 2015).

As innovative methods for integrating feedstock production into a wider landscape emerge, sustainability assessments are also evolving in the scientific literature for bioproducts and biorefinery concepts (eg, Heijungs et al., 2010; SPI, 2012; Horta Nogueira et al., 2015; Keller et al., 2015). Nevertheless, few practical cases exist and the majority of studies are conducted either on theoretical cases, modeling data, prefeasibility studies, or small pilot plants (eg, BIOCORE, 2014).

Often, results of sustainability assessments are not easy to understand. Different approaches leading to different results may confuse decisionmakers including policymakers, industry, and consumers. The opportunities and risks of increased biomass use must be weighed on the basis of clear and objective criteria and parameters. This is also true for the assessment of biorefinery concepts. If sustainability is not guaranteed, the acceptance of bioproducts and bioenergy will be at risk. Therefore, direct and innovative forms to communicate results of sustainability assessments should be considered in the future.

Science-based sustainability assessment can provide the basis for political decisionmaking in relevant areas, but decisionmaking also requires normative elements to deal with tradeoffs between economy and environment aspects of intergenerational equity (Heijungs et al., 2010). Addressing global and local issues simultaneously will require a global multistakeholder partnership, as well as partnerships at multiple levels of governance, from community and national levels up to international levels. Integrating poverty, food security, fair business practices, and environment will require sector-specific and cross-cutting policies that recognize interactions, tradeoffs as well as synergies, and take externalities and earth's carrying capacity into account. Furthermore, considerations for climate change need to be considered to favor bioproducts in the future. That means a coherent policy across energy, water, and food security issues. For this purpose, integrated knowledge and decisionmaking across disciplinary boundaries and policies is required.

While definitions for various types of biofuels exist, there is no agreed definition on bioproducts. The definition varies with respect to their functionality or the selected feedstock. For example, there are biobased biodegradable and biobased nonbiodegradable products available to the consumer. Biodegradability can be achieved independent from the feedstock; renewable feedstocks do not ensure biodegradability. The same molecules (eg, ethylene, glycerol, succinic acid) can be derived from fossil-based feedstock or biomass (see chapter: Development of Second-Generation Biorefineries). Hence, the environmental functionality can hardly be used as criteria for bioproducts. The preferred criterion for bioproducts is the biogenic content, but there is a need to clarify the referred molecule(s), for example, C, N, H, O, or their combination. Carbon is probably the most relevant molecule, although other biobased molecules should not be ignored in the scientific debate. However, there is no agreement yet on the proportion of biogenic carbon in a bioproduct.

In addition, the potential future penetration of the EU market with products made from renewable raw materials is still not well researched today. Two issues considered to influence how to reach the market are their ecological advantage (eg, recyclability) and cost. With respective policies in place, it is expected that the activities in biorefineries could stimulate the market. High-technology manufacturing (pharmaceutical) and medium-high-technology manufacturing (chemicals) have increased in recent years (US-DOE, 2011; BIOCORE, 2014). Green procurement is also expected to contribute to enhance the low-carbon economy (eg, USDA BioPreferred, 2015). Financial incentives may be possible through green procurement enforcement (Diaz-Chavez, 2014). Another key challenge is related to how to reach a broader market when significantly higher prices of bioproducts prevail compared to conventional (fossil-based) products. A further obstacle is the lack of information held by industrial and private consumers on the advantages of the latest products made from renewable raw materials (Oertel, 2007).

An EU standard is under development for assessing the sustainability of bioproducts, but it is not yet clear if the application will be reflected on an international level. Standards for bioproducts are not ready and a list of products (available and under research) needs to be created. Some measures need to be considered for bioproducts that are or will be in contact with food such as bioplastics used for packaging. Issues to consider include: biobased content, health, safety, environmental effects, and waste (Diaz-Chavez, 2014). Several EU policy instruments and green papers are related to the promotion of bioeconomy (eg, the Bioeconomy Strategy (EC, 2012);

Circular Economy Plan (EC, 2015); EU ecolabel (EU Ecolabel, 2015; European Bioplastics, 2014; EMAS, 2015), including packaging, monitoring of environmental impacts, and eco-labeling. Nevertheless, it has been reported (eg, LMI, 2009) that the different legal and regulatory instruments apply at different levels and this makes it difficult to influence all levels of the supply chain (manufacture, sale, and disposal of bioproducts) mainly because they are not one uniform product group, but a wide range of products with completely different characteristics, qualities, and uses. Some recommendations in the literature and interviews conducted demonstrate that this may create new opportunities and areas of R&D.

4.6 CONCLUSIONS AND RECOMMENDATIONS

Lessons learned from the production and use of first-generation biofuels have been considered in the development of sustainability assessment frameworks in general. At the same time, many frameworks primarily focus on feedstock production and are limited with respect to other areas, including health, safety, and final products as well as consumer behavior. Methodologies need further development, especially when compared to other fossil fuel origin products and their impact in the circular economy. Some of the main challenges for sustainability assessments of biobased products and biorefineries include:

- Finding a clear understanding of steps that can be taken to quantify and enhance the sustainability of bioproducts;
- Developing a consistent and transparent framework or minimum applicable criteria and indicators for sustainability assessment of bioproducts that are transparent and can be adapted to regional-specific conditions;
- Finding a consistent assessment approach for the provision of biomass for biobased products as well as bioenergy in general, and biofuels in particular;
- Lack of relevant data for all dimensions of sustainability collected in a consistent manner at different levels (from process level to national level);
- Costs and time involved to conduct a complete sustainability assessment in a consistent manner.

Also, with respect to good governance and policy assessment, there is a need to identify and foster the link between science, policies, and the decisionmaking process. Countries need to design, reform, and implement policies that value natural assets and align incentives with policy goals that promote sustainability of biobased systems.

Some general recommendations for future sustainability assessment are as follows:

- Prior to establishing a biorefinery, there must be a proper feasibility study along with a social and environmental impact assessment including all stakeholders that takes into account the whole supply chain.
- Further research needs to be conducted at the local level, including the participation of local stakeholders.
- Recommendations should be provided to international standards on how sustainability issues should be treated to facilitate trade of biomass feedstocks and biobased materials, as a "mature" bioeconomy will include exports and imports of both feedstocks and products.
- Standards need to be developed and applied for future bioproducts that are new to the market.
- Sustainability assessment research should consider additional issues such as health and safety in the bioeconomy.

REFERENCES

Babcock, B.A., 2015. Extensive and intensive agricultural supply response. Ann. Rev. Resour. Econ. 2015 (7), 333.

Batidzirai, B., 2013. Design of Sustainable Biomass Supply Chains. Optimising the Supply Logistics and Use of Biomass Over Time. Copernicus Institute. Utrecht University, The Netherlands.

Biobased Economy, 2015. CEN/TC 411 Bio-Based Products. Available from: <http://www.biobasedeconomy.eu/standardisation/cen-tc411/> (accessed November 2015.).

Bioconsortium, 2015. Bioeconomy: Circular by Nature Loop. The European Files. Available from: <http://biconsortium.eu/sites/biconsortium.eu/files/downloads/European_Files_september2015_38.pdf> (accessed November 2015).

BIOCORE Project, 2014. Integrated Assessment of Overall Sustainability. BIOCORE EC Funded Project. Available from: <http://www.biocore-europe.org/> (accessed November 2015).

Boddiger, D., 2007. Boosting biofuel crops could threaten food security. Lancet 370, 923–924.

Buchholz, T., Prisley, S., Marland, G., et al., 2014. Uncertainty in projecting GHG emissions from bioenergy. Nat. Clim. Chan. 4, 1045–1047.

Buchholz, T., Hurteau, M.D., Gunn, J., Saah, D., 2015. A global meta-analysis of forest bioenergy greenhouse gas emission accounting studies. GCB Bioenerg. n/a-n/a.

Buchholz, T.S., Volk, T.A., Luzadis, V.A., 2007. A participatory systems approach to modelling social, economic, and ecological components of bioenergy. Energ. Policy 35 (12), 6084–6094.

Cherubini, F., Bird, N.D., Cowie, A., Jungmeier, G., Schlamadinger, B., Woess-Gallasch, S., 2009. Energy- and greenhouse gas-based LCA of biofuel and bioenergy systems: key issues, ranges and recommendations. Resour. Conserv. Recy 53 (8), 434–447.

Chum, H.L., Zhang, Y., Hill, J., Tiffany, D.G., Morey, R.V., Eng, A.G., et al., 2013. Understanding the evolution of environmental and energy performance of the U.S. corn ethanol industry: evaluation of selected metrics. Biofuel. Bioprod. Bior. http://dx.doi.org/10.1002/bbb.1449.

Chum, H.L., Nigro, F.E.B., McCormick, R., Beckham, G.T., Seabra, J.E.A., Saddler, J., et al., 2015. Conversion technologies to biofuels and their use In: Souza, G.M.Victoria, R. Joly, C.Verdade, L. (Eds.), Bioenergy & Sustainability: Bridging the Gaps, vol. 72 SCOPE, Paris, pp. 368–461. Chapter 12, ISBN 978-2-9545557-0-6. Available from: <http://bioenfapesp.org/scopebioenergy/>.

Cibin, R., Chaubey, I., 2015. A computationally efficient approach for watershed scale spatial optimization. Environ. Model. Softw. 66, 1–11. http://dx.doi.org/10.1016/j.envsoft.2014.12.014.

Dale, V.H., Efroymson, R.A., Kline, K.L., Langholtz, M.H., Leiby, P.N., Oladosu, G.A., et al., 2013. Indicators for assessing socioeconomic sustainability of bioenergy systems: a short list of practical measures. Ecol. Indic. 26, 87–102.

Dale, V.H., Efroymson, R.A., Kline, K.L., Davitt, M., 2015. A framework for selecting indicators of bioenergy sustainability. Biofuel. Bioprod. Bior. 9 (4), 435–446. http://dx.doi.org/10.1002/bbb.1562.

Dale, V.H., Kline, K.L., Buford, M.A., Volk, T.A., Smith, C.T., Stupak, I., 2016. Incorporating bioenergy into sustainable landscape designs. Renew. Sustain. Energ. Rev. 56, 1158–1171. http://dx.doi.org/10.1016/j.rser.2015.12.038.

Dewulf, J., Van Langenhove, H. (Eds.), 2006. Renewables-Based Technology. Sustainability Assessment John Wiley and Sons, Ltd, USA.

Diaz-Chavez, R., 2010. Chapter 5: the role of biofuels in promoting rural development. In: Rosillo-Calle, F., Johnson, F. (Eds.), Food Versus Fuel: An Informed Introduction Zed Books.

Diaz-Chavez, R., 2014. D7.4: Final Assessment of the Social/Legal/Political Sustainability of the BIOCORE Biorefining System. BIOCORE EC funded project. Available from: <http://www.biocore-europe.org/>.

Diaz-Chavez, R., 2015. Assessing sustainability for biomass energy production and use. In: Rosillo-Calle, F., de Groot, P. (Eds.), The Biomass Assessment Handbook: Energy for a Sustainable Environment, 2nd ed. Earthscan, UK, 59–112.

EC, 2009. Directive 2009/28/EC of the European Parliament and of the Council of 23 April 2009 on the promotion of the use of energy from renewable sources and amending and subsequently repealing Directives 2001/77/EC and 2003/30/EC. Official Journal of the European Union; L 140/16 - L 140/62.

EC, 2012. Innovating for Sustainable Growth: A Bioeconomy for Europe. Communication From the Commission to the European Parliament, the Council, the European Economic and Social Committee and the Committee of the Regions. Brussels, 13.2.2012 COM (2012). Available from: <http://eur-lex.europa.eu/legal-content/EN/TXT/PDF/?uri=CELEX:52012DC0060&from=EN> (accessed December 2015.).

EC, 2015. Closing the Loop—An EU Action Plan for the Circular Economy. COM (2015) 614/2. Brussels. Available from: <http://ec.europa.eu/priorities/jobs-growth-investment/circular-economy/docs/communication-action-plan-for-circular-economy_en.pdf> (accessed December 2015.).

EMAS, 2015. The EU Eco-Management and Audit Scheme. Available from: <http://ec.europa.eu/environment/emas/index_en.htm> (accessed December 2015.).

EU Ecolabel, 2015. The EU Ecolabel. Available from: <http://ec.europa.eu/environment/ecolabel/> (accessed December 2015.).

European Bioplastics, 2014. Position Paper On. Available from: <http://en.european-bioplastics.org/wp-content/uploads/2014/publications/EuBP_PP_Assessment_of_the_sustainability_of_biobased_plastics.pdf> (accessed November 2015.).

FAO, 2011. Core Indicators on Bioenergy and Food Security. Food and Agricultural Organisation of the United Nations (FAO), Rome, Italy, see also: http://www.fao.org/energy/befs/en/.

Fritsche, U., Iriarte, L., 2014. Sustainability criteria and indicators for the bio-based economy in Europe: state of discussion and way forward. Energies 7 (11), 6825–6836.

GBEP, 2010. Analytical Tools to Assess and Unlock Sustainable Bioenergy Potential. The Global Bioenergy Partnership. FAO, Rome.

GBEP, 2011. The Global Bioenergy Partnership Sustainability Indicators for Bioenergy, 1st ed. FAO, Rome.

German Government, T.G.F., 2012. Biorefineries Roadmap. Federal Ministry of Food, Agriculture and Consumer Protection (BMELV), Berlin.

Heijungs, R., Huppes, G., Guinée, J.B., 2010. Life cycle assessment and sustainability analysis of products, materials and technologies. Toward a scientific framework for sustainability life cycle analysis. Polym. Degrad. Stabil. 95 (3), 422–428.

Helin, T., Sokka, L., Soimakallio, S., et al., 2013. Approaches for inclusion of forest carbon in life cycle assessment – a review. GCB Bioenerg. 5, 475–486.

Horta Nogueira, L.A., Leal, M.R.L.V., Fernandes, E., Chum, H.L., Diaz-Chavez, R., Endres, J., et al., 2015. Sustainable development and innovation In: Souza, G.M. Victoria, R. Joly, C. Verdade, L. (Eds.), Bioenergy & Sustainability: Bridging the Gaps, vol. 72 SCOPE, Paris, pp. 184–217. Chapter 6, ISBN 978-2-9545557-0-6. Available from: <http://bioenfapesp.org/scopebioenergy/>.

IEA-Bioenergy, 2015. Mobilizing Sustainable Bioenergy Supply Chains. Paris, France, Strategic Inter-Task study, commissioned by IEA Bioenergy and carried out with cooperation between IEA Bioenergy Tasks 37, 38, 39, 40, 42, and 43. Available from: <http://www.ieabioenergy.com/wp-content/uploads/2015/11/IEA-Bioenergy-inter-task-project-synthesis-report-mobilizing-sustainable-bioenergy-supply-chains-28ot2015.pdf> (accessed 9.12.15.).

International Food Policy Research Institute (IFPRI), 2015. Biofuels and Food Security Interactions Workshop Website (Online). Available from: <http://www.ifpri.org/event/workshop-biofuels-and-food-security-interactions> (accessed 10.2.15.).

Iqbal, J., Parkin, T.B., Helmers, M.J., Zhou, X., Castellano, M.J., 2014. Denitrification and nitrous oxide emissions in annual croplands, perennial grass buffers, and restored perennial grasslands. Soil Sci. Soc. Am. J. http://dx.doi.org/10.2136/sssaj2014.05.0221.

ISO, 2015. Sustainability criteria for bioenergy. ISO 13065:2015. International Standards Organisation Available from: <http://www.iso.org/iso/home/store/catalogue_tc/catalogue_detail.htm?csnumber=52528> (accessed November 2015.).

Johnson, W.P., Frederick, L.E., Millington, M.R., Vala, D., Reese, B.K., Freedman, D.R., et al., 2015. Potential impacts to perennial springs from tar sand mining, processing, and disposal on the Tavaputs Plateau, Utah, USA. Sci. Total Environ. 532, 20–30.

Jungmeier G. (2014). The Biorefinery Fact Sheet. Version 1.0, 2014-09-19. Available from: <http://www.iea-bioenergy.task42-biorefineries.com/upload_mm/d/f/7/445bc77c-5b98-4234-916e-59cd53d8f15b_8%20Biorefinery%20Fact%20Sheets%20IEA%20Task%2042%20170914.pdf> (accessed November 2015.).

Keam, S., McCormick, N., 2008. Implementing Sustainable Bioenergy Production; A Compilation of Tools and Approaches. IUCN, Gland, Switzerland. 32.

Keller, H., Rettenmaier, N., Reinhardt, G.A., 2015. Integrated life cycle sustainability assessment—a practical approach applied to biorefineries. Appl. Energ. 154 (2015), 1072–1081.

LMI, 2009. Taking Biobased From Promise to Market. A report from the Ad-hoc Advisory Group for Bio-based Product in the framework of the European Commission's Lead Market Initiative. Available from: <https://biobs.jrc.ec.europa.eu/sites/default/files/generated/files/policy/2009%20Report%20LMI%20Advisory%20group%20Bio-Based%20Products.pdf> (accessed October 2015.).

Luchner, S., Johnson, K., Lindauer, A., McKinnon, T., Max Broad, M., 2013. Social Aspects of Bioenergy Sustainability. U.S. Department of Energy, Washington D.C.

Macedo, I.C., Nassar, A.M., Cowie, A.L., Seabraa, J.E.A., Marelli, L., Otto, M., et al., 2015. Greenhouse gas emissions from bioenergy In: Souza, G.M. Victoria, R. Joly, C. Verdade, L. (Eds.), Bioenergy & Sustainability: Bridging the Gaps, vol. 72 SCOPE, Paris, pp. 582–617. Chapter 17, ISBN 978-2-9545557-0-6. Available from: <http://bioenfapesp. org/scopebioenergy/>.

Marland, G., Buchholz, T., Kowalczyk, T., 2013. Accounting for carbon dioxide emissions. J. Ind. Ecol. 17, 340–342.

Martins, A.A., Mata, T.M., Costa, C.A.V., Sikdar, S.K., 2007. A framework for sustainability metrics. Ind. Eng. Chem. Res. 46 (16), 5468.

McManus, M.C., Taylor, C.M., Mohr, A., Whittaker, C., Scown, C.D., Borrion, A.L., et al., 2015. Challenge clusters facing LCA in environmental decision-making—what we can learn from biofuels. Int. J. Life Cycle Ass. 20 (10), 1399–1414.

Miner, R.A., Abt, R.C., Bowyer, J.L., Buford, M.A., Malmsheimer, R.W., O'Laughlin, J., et al., 2014. Forest carbon accounting considerations in US bioenergy policy. J. For. 112 (6), 591–606.

Mohr, A., Raman, S., 2013. Lessons from first generation biofuels and implications for the sustainability appraisal of second generation biofuels. Energ. Policy 63 (2013), 114–122.

Muth, D.J., McCorkle, D.S., Koch, J.B., Bryden, K.M., 2012. Modelling Sustainable Agricultural Residue Removal at the Subfield Scale. Journal of Agronomy 104, 970–981.

Muth, D.J., Bryden, K.M., Nelson, R.G., 2013. Sustainable agricultural residue removal for bioenergy: a spatially comprehensive US national assessment. Appl. Energ. 102, 403–417.

O'Connell, D., Hatfield-Dodds, S., Braid, A., Cowie, A., Littleboy, A., Wiedmann, T., Clark, M., Raison, J., 2013. Designing for Action: Principles of Effective Sustainability Measurement. World Economic Forum. Available from: <http://www3.weforum.org/docs/GAC/2013/WEF_GAC_MeasuringSustainability__PrinciplesEffectiveSustainability Measurement_SummaryReport_2013.pdf>.

OECD, 2008. Conducting Sustainability Assessment. OECD Sustainable Development Studies. OECD, Paris, France.

OECD, 2010. Towards the Development of OECD Best Practices for Assessing the Sustainability of Bio-Based Products. OECD, Paris.30.

Oertel, D., 2007. Industrielle stoffliche Nutzung nachwachsender Rohstoffe. Sachstandsbericht zum Monitoring "Nachwachsende Rohstoffe" [The Industrial Use of Biomass. Report on the Monitoring of Biogenic Resources]. Arbeitsbericht, Berlin, Germany, Available from: <http://www.tab.fzk.de/de/projekt/zusammenfassung/ab114.pdf> (accessed January 2013.).

Panichelli, L., Gnansounou, E., 2015. Impact of agricultural-based biofuel production on greenhouse gas emissions from land-use change: key modelling choices. Renew. Sustain. Energ. Rev. 42, 344–360.

Parish, E.S., Kline, K.L., Dale, V.H., Efroymson, R.A., McBride, A.C., Johnson, T.L., et al., 2013. A multi-scale comparison of environmental effects from gasoline and ethanol production. Environ. Manage. 51 (2), 307–338. http://dx.doi.org/10.1007/s00267-012-9983-6.

Patel, A.D., Telalović, S., Bitter, J.H., Worrell, E., Patel, M.K., 2015. Analysis of sustainability metrics and application to the catalytic production of higher alcohols from ethanol. Catal. Today 239, 56–79.

Pinazo, J.M., Domine, M.E., Parvulescu, V., Petru, F., 2015. Sustainability metrics for succinic acid production: a comparison between biomass-based and petrochemical routes. Catal. Today 239, 17–24.

Plevin, R., Beckman, J., Golub, A.A., Witcover, J., O'Hare, M., 2015. Carbon accounting and economic model uncertainty of emissions from biofuels-induced land use change. Environ. Sci. Technol. 49 (5), 2656–2664.

Raman, S., Mohr, A., Helliwell, R., Ribeiro, B., Shortall, O., Smith, R., et al., 2015. Integrating social and value dimensions into sustainability assessment of lignocellulosic biofuels. Biomass Bioenerg. 82, 49–62. November 2015.

Rosillo-Calle, F., Johnson, F. (Eds.), 2010. Food Versus Fuel. An Informed Introduction to Biofuels Zed Books, UK.

Rosillo-Calle, F., de Groot, P., Hemstock, S.L., Woods, J. (Eds.), 2015. The Biomass Assessment Handbook: Energy for a Sustainable Environment, 2nd ed. Earthscan, UK.

Sacramento-Rivero, J., 2012. A methodology for evaluating the sustainability of biorefineries: framework and indicators. Biofuel Prod. Bior. 6 (1), 32–44.

Searchinger, T., Heimlich, R., 2015. Avoiding Bioenergy Competition for Food Crops and Land. Working Paper, Installment 9 of Creating a Sustainable Food Future. World Resources Institute, Washington, DC, Available from: <http://www.worldresourcesreport.org>.

Sherval, M., 2015. Canada's oil sands: the mark of a new 'oil age' or a potential threat to Arctic security? Extr. Ind. Soc. 2 (2), 225–236.

Sikdar, S.K., 2007. Sustainability perspective and chemistry-based technologies. Ind. Eng. Chem. Res. 46 (14), 4727–4733.

Souza, G.M., Victoria, R. Joly, C., Verdade, L. (Eds.), 2015. Bioenergy & Sustainability: Bridging the Gaps, vol. 72 SCOPE, Paris, pp. 779. ISBN 978-2-9545557-0-6, Available from: <http://bioenfapesp.org/scopebioenergy/>.

SPI, 2012. Understanding Biobased Carbon Content. Society of the Plastics Industry Bioplastics Council February 2012. Available from: <http://www.plasticsindustry.org/files/about/BPC/Understanding%20Biobased%20Content%20-%200212%20Date%20-%20FINAL.pdf> (accessed November 2015.).

Ssegane, H., Negri, M.C., Quinn, J., Urgun-Demirtas, M., 2015. Multifunctional landscapes: site characterization and field-scale design to incorporate biomass production into an agricultural system. Biomass. Bioenerg. 80, 179–190.

Taheripour, F., Tyner, W.E., 2013. Biofuels and land use change: applying recent evidence to model estimates. Appl. Sci. 3, 14–38.

Ter-Mikaelian, M.T., Colombo, S.J., Chen, J., 2015. The burning question: does forest bioenergy reduce carbon emissions? A review of common misconceptions about forest carbon accounting. J. For. 113 (1), 57–68.

UN, 2015. Sustainable Development Knowledge Platform. United Nations. Available from: <https://sustainabledevelopment.un.org/> (accessed December 2015.).

USDA BioPreferred, 2015. What Is BioPreferred? Available from: <http://www.biopreferred.gov/BioPreferred/faces/pages/AboutBioPreferred.xhtml> (accessed January 2016.).

U.S-DOE, 2011. Department of Energy. U.S. Billion-ton Update: Biomass Supply for a Bioenergy and Bioproducts Industry. R.D. Perlack and B.J. Stokes (Leads), ORNL/TM-2011/224. Oak Ridge National Laboratory, Oak Ridge, TN.227.

van Dam, J., Junginger, M., Faaij, A.P.C., 2010. From the global efforts on certification of bioenergy towards an integrated approach based on sustainable land use planning. Renew. Sustain. Energ. Rev. 14 (9), 2445–2472.

Voegele, E. (2015). EU Agriculture and Fisheries Council Adopts ILUC Biofuel Rule. Biodiesel Magazine. Available from: <http://www.biodieselmagazine.com/articles/449113/eu-agriculture-and-fisheries-council-adopts-iluc-biofuel-rules> (accessed November 2015.).

Warner, N.R., Christie, C.A., Jackson, R.B., Vengosh, A., 2013. Impacts of shale gas wastewater disposal on water quality in Western Pennsylvania. Environ. Sci. Technol. 47 (20), 11849–11857.

Youngs, H., Somerville, C., 2014. Best practices for biofuels—data-based standards should guide biofuel production. Science 344 (6188), 1095–1096.

CHAPTER 5

Biomass Supply and Trade Opportunities of Preprocessed Biomass for Power Generation

B. Batidzirai[1], M. Junginger[2], M. Klemm[3], F. Schipfer[4] and D. Thrän[3, 5]

[1]Energy Research Centre, University of Cape Town, Cape Town, South Africa
[2]Copernicus Institute, Utrecht University, Utrecht, The Netherlands
[3]German Biomass Research Center (DBFZ), Leipzig, Germany
[4]EEG, Vienna University of Technology, Vienna, Austria
[5]Department of Bioenergy, Helmholtz Center for Environmental Research (UFZ), Leipzig, Germany

Contents

Abstract

International trade of solid biomass is expected to increase significantly given the global distribution of biomass resources and anticipated expansion of bioenergy deployment in key global power markets. Given the unique characteristics of biomass, its long-distance trade requires optimized logistics to facilitate competitive delivery value chains. Preprocessing biomass via pelletizing, torrefaction, and hydrothermal carbonization potentially improves bioenergy supply economics as illustrated by two case studies in this chapter. The case studies presented in this chapter compare woody and herbaceous biomass value chains and demonstrate that it is feasible and desirable in current conditions to establish large-scale conversion plants close to mature electricity markets and source preprocessed biomass from the international market. In the short term, conventional pellets are expected to play an important role as the internationally traded solid biomass commodity and feedstock in biopower production. In the near future, torrefied pellets may become the dominant and preferred internationally traded solid biomass commodity as the technology is commercialized. Hydrothermal carbonization technology is also still under development, but has the potential to unlock additional feedstock from wet biomass streams. Successful deployment of these technologies is expected to improve bioenergy supply chains in terms of costs and greenhouse gas impacts. Local bioenergy markets are also expected to develop, and provide localized opportunities for local biomass production and use. Utilization of herbaceous biomass and agricultural residues for power production is a promising option, but its application in cofiring is yet to be proven on a wide commercial scale. The analysis of agricultural residue mobilization in South Africa demonstrates that preprocessing also plays a major role in improving biomass delivery costs and subsequent electricity generation costs in local markets.

5.1 INTRODUCTION

Global demand and trade of solid biomass have been growing rapidly over the past decade, especially in the power and heat sectors (Lamers et al., 2012; Sikkema et al., 2011; Cocchi et al., 2012). This demand is driven mainly by renewable energy targets and incentives (Goh et al., 2012; Lamers et al., 2012), as well as energy security and environmental objectives (Chum et al., 2011; Beekes and Cremers, 2012; Tarcon, 2011). It is anticipated that bioenergy use will grow considerably in the near future. The International Energy Agency (IEA) estimates that biomass will contribute about 71.5 EJ to total global energy supply (under a Current Policies Scenario) by 2035 (OECD/IEA, 2011) and biomass power contribution is expected to increase to about 18% by 2050 (under the Blue Map scenario) (OECD/IEA, 2010). European Commission assessments (EC, 2014a) also project further increases in biomass use in the heat and power sector as the European Union (EU) implements a transition to a low-carbon economy by 2050. Biomass-based electricity is expected to grow to between 336 and 520 TWh by 2050 with installed capacities of between 39 and 66 GW in the same period in the EU alone (EC, 2014b).

5.1.1 Biomass Supply and Demand Centers

Key biomass power markets are currently centered around and likely to remain in Europe, North America, and East Asia as shown in Fig. 5.1. However, apart from North America, these regions have limited available biomass resources to meet current and projected future biomass demand. Major global biomass production regions are located in North America, Russia, Scandinavia, South America, and parts of Africa and Asia

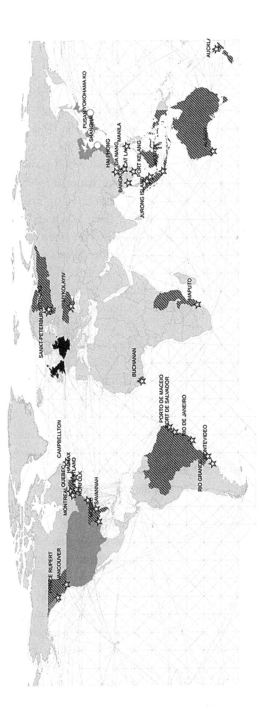

Figure 5.1 Global distribution of key solid biomass demand (solid, circle harbors) and supply regions (shaded, starred harbors) with respective harbors. *From Lamers, P., Hoefnagels, R., Junginger, M., Hamelinck, C., Faaij, A., 2015. Global solid biomass trade for energy by 2020: an assessment of potential import streams and supply costs to North-West Europe under different sustainability constraints. GCB Bioenerg. 7 (4), 618–634.*

(Goh et al., 2012; Smeets et al., 2007). Generally, countries with large biomass resource potentials have vast territories, and therefore the resources are often dispersed in a large territory and difficult to access, such as Russia and Canada.

Given the global spatial distribution of biomass and anticipated future expansion of bioenergy deployment in key global markets, substantial increases in international bioenergy trade are inevitable (Matzenberger et al., 2015). This transition to the large-scale commodification of solid biomass started a decade ago (Chum et al., 2011) and is still growing rapidly (Lamers et al., 2014). Already a burgeoning international solid biomass trade is evident, 18 Mt of solid biomass were traded in 2010 up from 3.5 Mt in 2000 (of which wood pellets trade increased from 0.5 to 6.6 Mt over the same period) (Lamers et al., 2012; Tustin, 2012). This large-scale international biomass trade is mainly linked to economic drivers and regional biomass availability (Faaij and Domac, 2006). The growth in biomass trade will assist to develop and maintain international bioenergy markets as well as develop currently underutilized bioenergy potentials in many regions of the world (Faaij et al., 2014).

To meet the growing global biomass demand and to mobilize these large-scale biomass supplies, large volumes of biomass feedstock need to be secured, and competitive feedstock value chains need to be developed and optimized, based on identification of appropriate combinations of feedstock and preprocessing technologies (Batidzirai et al., 2013).

There are two main types of solid biomass feedstocks of interest for international traded solid biomass—woody and herbaceous biomass. These two biomass types have distinct differences in their characteristics and value chains, which ultimately impact the competitiveness of delivered biomass (Batidzirai et al., 2014). Herbaceous biomass includes grasses such as switchgrass and miscanthus as well as agricultural residues such as wheat straw and corn stover. Its markets are much less mature and its international trade is currently limited due to various challenges associated with the large-scale logistics and conversion into energy or products. Compared to woody biomass, the lower heating value and lower bulk density of herbaceous biomass bales ($100–140\,kg\,m^{-3}$) and corresponding pellets, result in much higher transportation costs per unit delivered energy than for woody biomass. Also, the higher sulfur and chlorine content can lead to fouling of equipment and require changes in process design; this increases conversion costs. Herbaceous biomass is also problematic as it ignites easily, posing storage difficulties. There is also limited experience in preprocessing herbaceous biomass. However, herbaceous biomass sourcing costs are typically lower than woody biomass (Batidzirai et al., 2013, 2014). It is therefore important to evaluate the implications of international supply of these two types of biomass feedstocks. Whereas woody biomass pellets are already mature and a "flowable commodity" in the power and heat market, there are uncertainties around the competitive supply of preprocessed herbaceous biomass.

This chapter assesses the opportunities for regional and international biomass supply and trade of preprocessed biomass primarily for power generation purposes. Based on two case studies, the chapter compares the performance of various biomass supply chain configurations, based on different preprocessing technologies, types of biomass feedstocks, and biopower markets.

5.2 INTERNATIONAL TRADE AND SUPPLY OPPORTUNITIES OF PROCESSED STABLE BIOMASS INTERMEDIATES FOR BIOPOWER MARKET

5.2.1 Development of Biopower Markets

As the major biomass market, solid biomass in the EU is mainly used for heating (~85%) and electricity generation (~15%). Over 90% of this biomass is domestically produced in the EU and is used mainly for household and other small-scale heating applications (EC, 2014b). The large-scale solid biomass requirements (such as for cofiring) are increasingly imported from outside the EU. There is a significant market for cofiring preprocessed biomass (predominantly woody biomass) with coal in power generation, especially in northern and western European countries. These markets are driven mainly by the availability of feed-in premiums or quotas for green electricity, and other government policies. The key biomass feedstock has been industrial wood pellets imported mainly from Canada, the United States, and Russia (Goh et al., 2012). Intra-EU solid biomass trade, for example, from the Baltic states to Sweden and Denmark or from Austria to Italy contributed about two-thirds of cross-border trade by 2010 (Lamers et al., 2012). Prospects for market growth in biomass cofiring power generation are positive, and over the past decade, there has been an increase in pellet demand for cofiring in Belgium, the Netherlands, the United Kingdom, and Denmark, mainly driven by government policies. Significant growth is projected in solid biomass-based cofiring in the EU, more than doubling from about 74TWh in 2012 to 157TWh in 2020. According to EC (2014b), wood pellet imports to the EU are set to increase from 4.3 million t[1] in 2013 to 15–30 million t by 2020 to meet the expected demand for large-scale cofiring and combined heat and power (CHP) applications. Thus the EU is likely to remain the key driver of solid biomass trade specifically targeting the power sector.

East Asia, particularly Japan and South Korea, have also set renewable energy targets, which have stimulated cofiring of wood pellets in large coal power plants. Both countries are expected to experience strong growth in consumption in the next few years (Goh et al., 2012).

5.2.2 The Importance of Preprocessing

International trade of biomass over long distances is costly and can render biomass uncompetitive (Rentizelas et al., 2009; Hamelinck et al., 2005). This is because biomass has unique characteristics that necessitate preprocessing before it can be efficiently stored, transported, or used in various applications currently designed for fossil fuels (Tumuluru et al., 2012). Biomass is often available seasonally in small quantities scattered over many locations (Junginger et al., 2001; Deng et al., 2009). It is highly heterogeneous, which results in wide variations in combustion properties (Tapasvi et al., 2012). It usually has a high moisture content and consequently low heating value (Ben and Ragauskas, 2012). It is hydrophilic and biodegradable, posing storage problems (Tumuluru et al., 2012). Its combustion efficiency is lower than fossil fuels (Crocker and Andrews, 2010), which decreases the capacity of given systems. Biomass therefore often needs to be pretreated to improve its characteristics and associated handling (Rentizelas et al., 2009; Luo, 2011). However, preprocessing costs are significant and can render biomass uneconomical (EverGreen, 2009).

Biomass preprocessing includes baling or bundling (for agricultural and forestry residues), sizing (into chips or flour, for example), drying, torrefying, and densification into conventional pellets (CPs), briquettes, or torrefied pellets (TOPs). Hydrothermal carbonization (HTC) is also another preprocessing technology especially suitable for conditioning wet biomass streams. Hence, an important logistical question is to identify combination(s) of preprocessing options which can best upgrade biomass properties for optimal downstream logistics.

---

[1] 1 metric tonne = 1000 kg = 1 Megagram (Mg).

5.2.2.1 Pelleting and Torrefaction

Pelleting biomass is currently the most important preprocessing approach for solid biomass, and wood pellets are currently the most important internationally traded biomass commodity (Lamers et al., 2012). The technology is mature and markets have developed in the power and heat sectors (Chum et al., 2011; Goh et al., 2012). Although it is yet to be proven on a commercial scale, torrefied pellets appear to have more advantages compared to CPs (Batidzirai et al., 2013). TOPs have a higher energy density ($12–20\,GJ_{LHV}\ m^{-3}$ compared to $7–10.4\,GJ_{LHV}\ m^{-3}$ for conventional pellets) (Bagramov, 2010; Tumuluru et al., 2012; Kiel et al., 2012; Melin, 2011; Boyd et al., 2011) and this has potential to lower logistic costs.

Torrefaction (combined with pelletization) is a promising biomass preprocessing technology which has potential to produce a homogeneous biomass carrier with improved energy density and combustion characteristics, and whose properties closely match those of low-grade coal (Agar and Wihersaari, 2012; Li et al., 2012; Phanphanich and Mani, 2011). This would allow cofiring with higher percentages of biomass than is currently possible with conventional pellets (Beekes and Cremers, 2012; Meerman et al., 2012).

Given the global distribution of biomass production regions and key markets (Chum et al., 2011), preprocessing biomass plays an important role in improving biomass supply chain economics, and enables biomass to be delivered to the market cost-effectively with lower downstream investments (Uslu et al., 2008; Miao et al., 2012; Bergman, 2005). This would also allow access to remote biomass resources and improve the potential of biomass as a renewable energy resource.

5.2.2.2 Hydrothermal Carbonization

Hydrothermal carbonization (HTC) enables the conversion of especially wet biomasses into a solid fuel—so-called HTC coal (Sevilla, 2009; Libra et al., 2011; Dinjus et al., 2011; Funke and Ziegler, 2010). Beside biomass types with established applications in combustion or biogas production, there is potential for harnessing wet and hardly biodegradable biomass like food industry waste, municipal biowaste, digestates from biogas production processes, and sewage sludge (Escala et al., 2013). The utilization of these wet biomass resources is of major importance for the expansion of bioenergy feedstock base (Wilén et al., 2013; Statistisches Bundesamt, 2013). Other possibilities of thermo-chemical conversion of wet biomass are very limited because of the energy demand for drying. Because the reaction medium is water, wet biomass does not need to be dried. During HTC, the biomass or waste is converted with water as reaction agent at 180–250°C and 10–40 bar. Currently, typical HTC process operational times are between 1.5 and 6 hours.

In comparison with the input wet biomass, there are improvements in major properties of HTC coal such as heating value, carbon content, volatile mater, homogeneity, and defined structure. A biomass quality close to lignite coal can be reached (Ramke et al., 2010; Kietzmann et al., 2013; Clemens et al., 2012; DBFZ, 2013). Different energy applications are possible, especially as a coal substitute (Tremel et al., 2012; Gunarathne et al., 2014). The dewatering of this coal can reach a high dry matter content and that is why energy demand for coal drying is low. In particular because of this, HTC can be the energy-efficient alternative in many cases.

Many different HTC plant concepts have been developed, mostly in Germany but also in other countries (Hitzl et al., 2014; Artec, 2015; AVA-CO2, 2015; CS CarbonsSolutions, 2015; SunCoal, 2015; TerraNova, 2015). These plants are for demonstration purposes (Klemm et al., 2015), and none is currently in commercial operation. Thus, the development of an HTC coal market is still in its early stages.

5.2.3 Location of Final Conversion Facility

The strategic location of the final biomass conversion plant (as well as preprocessing facilities) is an important consideration for the competitive utilization of biomass. Given the capital-intensive nature of conversion plants, effective use should be made of economies of scale and centralized/decentralized processing where appropriate. Typically, preprocessing can be cost-effectively achieved in decentralized small-scale operations where facilities are located near the plantation or source of biomass, which helps to reduce logistic capacity very early in the chain. However, there is a trade-off between the size of the preprocessing plant and raw biomass transportation costs, which are also affected by the availability and distribution of sufficient feedstock in the vicinity of the processing plant. Batidzirai et al. (2014) established that at current technology costs, the optimal plant size for pellet plants is around 250,000 t year^{-1}. However the trade-off between transport costs of biomass supply and unit scaling effects has to be calculated for every plant individually since multiple parameters can be decisive for this optimization. In Schipfer et al. (2015) the combination of feedstock yield, its availability, and accessibility are outlined to be critical as well as earlier inflexion points for CPs than for TOPs.

Where local economics are attractive, final conversion in the biomass production regions (early in the supply chain) can be beneficial and cost-effective. However, establishing large-scale conversion facilities in major biomass feedstock production regions (typically in developing countries) involves significant technological and commercial risk, and capital costs may be higher (Batidzirai et al., 2014). According to IRENA (2012), financiers consider biomass power projects to be risky, as existing projects in developing countries have failed to meet expected performance. Economies of scale play an important role in driving down production costs and such large-scale conversion facilities are more suited for well-developed bioenergy markets where policy measures favor their establishment. For some regions, biobased power generation may not be competitive against established technologies such as hydro. An important consideration for locating the final conversion plants is the availability of well-developed infrastructure in the importing country, such as deep harbors with storage capacity to handle large volumes of biomass imports from different countries.

In the short to medium term, western Europe is likely to remain the main market for internationally traded solid biomass and ideal location for establishing biobased power generation facilities given the regional drive to increase the share of renewables to 20% by 2020 (EC, 2014b). To enable the transition to a low-carbon economy, significant investment in renewable energy electricity is inevitable. Given the projected contribution of biomass to future electricity mix in the EU, growth in solid biomass imports from the international market is a key strategy for many EU countries. Major utilities in the region, such as RWE-Essent, Vattenfall, Dong Energy, Drax, GDF Suez, and Eon, are already actively pursuing biomass cofiring strategies and importing millions of tonnes of solid biomass every year (Verhoest and Ryckmans, 2014). Stakeholder consultations in the EU have shown that trade is essential for reliability of supply of biomass and offer flexibility as sourcing biomass from different regions reduces the feedstock supply risks (EC, 2014b). The case studies presented below demonstrate that it is feasible and desirable at current conditions to establish final conversion plants close to the electricity market and source biomass feedstock from the international market.

5.2.4 Energy Crop-Based Supply Chains: Mozambique Case Study

To demonstrate the competitiveness of international supply of preprocessed biomass for cofiring in the power sector, we present a case study that compares the supply of conventional pellets (CPs) and torrefied pellets (TOPs) from southeast Africa and subsequently use in cofired power plants in western Europe, taking Rotterdam as a location of the final conversion facility. First, we compare the economic performance of TOPs and CPs based on different feedstocks (eucalyptus and switchgrass). Second, the study evaluates the impact of supplying biomass from different regions (productive and marginal land quality in Mozambique, Nampula, and Gaza, respectively). Third, we compare dedicated biomass-fired power generation (BtP[2] or TtP) and cofiring biomass with coal in a coal-biomass to power (CBtP or CTtP) plants. Lastly, the study compares the competitiveness of supplying different markets (close to biomass production sites in southern Africa or in major international bioenergy markets in the Netherlands). This comparison is performed for both the short term (current) and long term (2030). Costs are given in US$$_{2010}$.

In this case study, biomass feedstock (eucalyptus and switchgrass) is produced in Mozambique, and undergoes preprocessing before shipment to Europe for power production. A comparison is also made for the local conversion of biomass in Mozambique. Key assumptions include interest rates of 8% (international), 13% for Mozambique (Trading Economics, 2014) and exchange rate of 1.30 US$/€, 30 Mozambican Meticais/US$. Further assumptions and input data are available in Batidzirai et al. (2014).

Figs. 5.2 and 5.3 show the economic performance of different supply chains considered in this case study, for the short and long term, respectively.

It is apparent that feedstock, truck transport, and conversion costs are dominant and together constitute up to 90% of final, delivered electricity costs. Conversion is the most important cost element and represents up to 56% of overall power supply costs. Preprocessing costs are also important, contributing up to 20% to the final electricity costs. Also, lower-cost electricity is produced from chains based on the more productive Nampula region ($81–107 GJ^{-1}) compared to Gaza ($85–107 GJ^{-1}). It is also clear that electricity from supply chains based on switchgrass is produced at lower cost than from eucalyptus for both the short term and long term. This is mainly attributed to the lower production costs of switchgrass. In the short term, biomass from switchgrass (both CPs and TOPs) is delivered at $5.1–7.3 GJ^{-1} compared to biomass from eucalyptus ($5.4–7.5 GJ^{-1}).

---

[2] Final conversion is denoted by XtP, where P is power, X is either coal (C), biomass/pellets (B), TOPs (T) or combinations such as CB for cofiring.

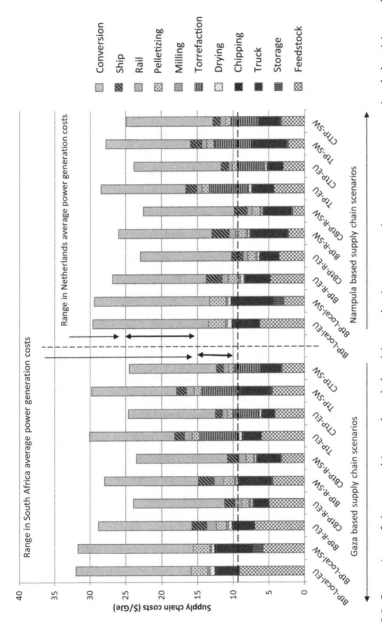

Figure 5.2 Comparison of short-term biomass-based electricity production costs against current average national electricity production costs in South Africa and Netherlands.

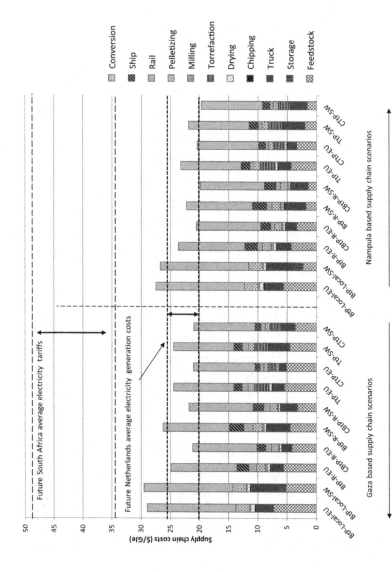

Figure 5.3 Comparison of long-term biomass-based electricity production costs against projected national electricity production costs in the Netherlands and electricity tariffs in South Africa.

In the long term, switchgrass is delivered at \$4.3–5.8 GJ^{-1}, while eucalyptus is delivered at \$4.8–5.8 GJ^{-1}. Due to the low bulk density of switchgrass bales (100–140 kg m^{-3}), truck transportation from the field costs for switchgrass are much higher (\$1.7–2.0 GJ^{-1}) than for eucalyptus (\$1.1–1.4 GJ^{-1}), as logs have a bulk density of around 460 kg m^{-3}. In addition, switchgrass incurs high storage costs at the farm, as it is more susceptible to moisture increases and dry matter losses if stored in the open.

Currently, torrefied pellets are delivered in Rotterdam at higher cost (\$6.5–7.5 GJ^{-1}) than conventional pellets (\$5.1–6.2 GJ^{-1}). In the long term, torrefied pellets are expected to decline in cost (\$4.7–5.8 GJ^{-1}) and converge with conventional pellets (\$4.3–5.3 GJ^{-1}). These differences are due to lower efficiency of torrefaction chains and the higher torrefaction production costs compared to CPs. Torrefaction is an additional costly step (at least \$2 GJ^{-1} in the short term) compared to conventional pelletizing (about \$0.5 GJ^{-1}). However, due to improved logistics and lower conversion investment requirements, electricity production costs from TOPs are lower than from CPs, especially in the long term. Final power production costs are influenced by the properties of the feedstock as raw biomass conversion requires additional investment and results in decreased plant capacity and efficiency. These additional costs are lower for TOPs than for CPs as TOPs have characteristics that are closer to coal. However, the additional costs of torrefying biomass do not offset the benefits of lower conversion costs as conventional pellet-based chains deliver lower cost electricity (\$81–115 GJ^{-1}) than the TOPs (\$86–108 GJ^{-1}), albeit marginally.

It is also apparent from these scenarios that cofiring biomass with coal results in lower electricity production costs (\$81.0–89.7 MWh^{-1}), compared to biomass-only fired power plants (\$93.5–108.4 MWh^{-1}), since new capital investment for retrofitting cofired power plants is much lower than when establishing greenfield power plants. Cofiring scenarios (CXtP)—based on switchgrass from productive land—have the best economics for both the short and long term (\$22.5 GJ^{-1} or \$81.0 MWh^{-1}). Although delivered biomass feedstock costs are much lower in Mozambique due to avoided international logistics, power production in Mozambique (\$106–115 MWh^{-1}) is more costly than in Rotterdam (\$81–108.4 MWh^{-1}). This is due to the relatively higher investment costs of smaller-scale plants (assumed for Mozambique) and higher interests rates (13%) compared to the Netherlands (8%). In addition, power production is not competitive in the Mozambican market, as these costs are much higher than the average levelized electricity generation costs for southern Africa (\$32–54 MWh^{-1}) (IEA, 2010).

For comparison, the average electricity tariffs in southern Africa are about $70 MWh^{-1} (NERSA (http://www.nersa.org.za)).

For the Netherlands, power production costs ($81–108.4 MWh^{-1}) are competitive against the average power generation costs in the Netherlands ($55–91 MWh^{-1}) (IEA, 2010); and much lower compared to average electricity tariffs in Netherlands (estimated to be $198 MWh^{-1}) (Europe Energy Portal, 2013).

Long-term power production costs across all scenarios are estimated to be 6–21% lower at $71–106 MWh^{-1} than in the short term. This is attributed to technological learning and scaling up of facilities across the bioenergy value chain, especially in critical components of feedstock production, preprocessing, and conversion. Conversion into final products dominates overall costs representing 42–57% of power production costs across all scenarios. Although feedstock costs are important, they account for a much lower proportion of total costs (7–25% of overall costs). Regional cost differences across the scenarios are marginal but evident; long-term power production in Gaza scenarios range from $76 MWh^{-1} to $106 MWh^{-1}, while in Nampula costs are $70–99 MWh^{-1}. Power production in Mozambique ($96–106 MWh^{-1}) is more costly than in Rotterdam ($71–95 MWh^{-1}). As shown in Fig. 5.3, the lowest cost power pathway ($19.7 GJ^{-1} or $71 MWh^{-1}) is associated with the cofiring switchgrass TOPs in Rotterdam. For the Netherlands, these future power production costs are competitive compared to future expected electricity generation costs ($70–90 MWh^{-1}) (van den Broek et al., 2011). However, future electricity tariffs in southern Africa[3] are expected to be much higher at $120–173 MWh^{-1} (DOE, 2011)—but these tariffs already include transmission, supply charges, and taxes.

5.3 LOCAL/REGIONAL TRADE AND SUPPLY OPPORTUNITIES OF RAW BIOMASS FOR BIOENERGY MARKET

Local and regional markets for biomass offer opportunities for developing the bioenergy sector in different parts of the world. Several countries

[3]We compare the power production costs in Mozambique with southern African tariffs (since power supplies in the region countries are intricately linked under the southern African power pool [SAPP]). Mozambique both exports and imports electricity from SAPP, its 2075 MW hydro plant supplies mainly South Africa while 850 MW are imported from South Africa to supply the southern region 136. In addition, future electricity prices are available for southern Africa based on South African projections and not for Mozambique.

have established small- to large-scale biomass-based power generation facilities, for example, Sweden, Germany, and the United Kingdom. Local and regional biomass markets are important especially for the utilization of biomass resources in regions without adequate infrastructure (large-scale preprocessing and logistical) for supplying large volumes of biomass to international markets. Forestry and agricultural residues are especially an important resource which can be sustainably harnessed and utilized locally with minimal preprocessing such as bundling and baling. Regional trade of such biomass feedstock allows small-scale biomass producers to add value to biomass and get additional income through diversification of their operations.

Several countries such as Denmark, the United Kingdom, Spain, Sweden, China, and India have developed large-scale crop residue to energy facilities (Peidong et al., 2009; Purohit, 2009; Urošević and Gvozdenac-Urošević, 2012). Key crop residues include corn stover, wheat straw, rice straw and husks, and bagasse (Chum et al., 2011; Perlack et al., 2005; Kline et al., 2008). Globally, the use of sugarcane bagasse for power and heat production is the most common and mature energy application of crop residues for those countries with large sugarcane industries (REN21, 2011). There is less experience in energy conversion for other crop residues, but interest is significant in using corn stover for advanced biofuels, especially in the United States (Tyndall et al., 2011; Chum et al., 2011; USDOE, 2012). In Europe, Denmark pioneered large-scale power generation using straw and has commercialized the technology since 1989 (Skøtt, 2011; Kretschmer et al., 2012).

According to IPCC biomass energy deployment scenarios (Chum et al., 2011), agricultural residues are likely to play an important role in future energy systems contributing between 15 and 70 EJ to the long-term global energy supply. Agricultural residues represent an important energy resource for countries with a large agricultural production base (WBGU, 2009; Chum et al., 2011; Dornburg et al., 2010). Although there is a large untapped potential for agricultural residues globally, there is little experience in their application for large-scale power production. Also due to the diversity of agricultural residues and differences in their chemical and physical characteristics, their utilization requires modifications in value chain and at the final conversion plant. Thus their local and regional application could allow the resource to benefit from technological learning and eventual deployment into the international market. We discuss below a case study conducted for South Africa to establish the feasibility of mobilizing agricultural residues for large-scale energy applications.

5.3.1 Agricultural Residues-Based Supply Chains: South Africa Case Study

This case study assesses the feasibility of mobilizing corn and wheat residues for large-scale power production in South Africa by establishing sustainable residue removal rates at the farm level and electricity production costs based on different biomass production regions at Camden (1600 MWe out), a depreciated power plant in Mpumalanga province. A key outcome of this case study was to estimate the national crop residue harvesting potential for bioenergy use, while maintaining soil productivity and avoiding displacement of competing residue uses. At every stage of the agricultural residues value chain, the study identified measures that would improve the performance of the overall crop residue supply chain and enhance the competitiveness of biomass- compared to fossil-based power generation. This included a comparison of applying different preprocessing technologies such as pelleting and torrefaction.

Currently, the sustainable bioenergy potential from corn and wheat residues is estimated to be about 6 million t (104 PJ), out of an annual gross crop residue potential of about 14.4 million t. This sustainable potential included 5.1 million t of corn stover and 600,000 t of wheat straw. About 4.2 million t of corn stover would be required for soil erosion control while 9.3 million t would be required for soil organic carbon (SOC) maintenance. Also, about 260,000 t of corn stover are required to meet cattle feed demand. Similarly, 870,000 t and 100,000 t of wheat straw are required to maintain SOC and prevent erosion, respectively. About 70,000 t of wheat straw are utilized as livestock bedding.

There is potential to increase the amount of crop residues to 238 PJ through measures such as no till cultivation and adopting better cropping systems. These estimates were based on minimum residue requirements of 2 t ha^{-1} for soil erosion control and additional residue amounts to maintain 2% SOC level.

5.3.1.1 Corn and Wheat Residue Costs at the Farm Gate

Overall the cost of collecting, baling, and storing corn stover at the farm is estimated to be about $1.5 GJ^{-1}. Compensation for the farmers for lost nutrients dominates the cost of corn stover at the farm accounting for 58% of total costs (or $0.87 GJ^{-1}). Baling is also a very important cost element representing 29% of the total costs. A 10% farmer profit margin on direct costs is allowed in the estimated direct costs and this also represents about 4% of the total costs. Wheat straw at a typical dryland farm costs about \$1.5 GJ^{-1} assuming a yield of 2t ha^{-1}. Baling dominates the overall wheat straw costs at 43% (or \$0.66 GJ^{-1}) and farmer nutrient compensation accounts for 41%.

Overall, about 7% of crop residues (6.8 PJ) are available at costs below \$1 GJ^{-1} at the farm gate while 34% of the residues are available at costs below \$1.2 GJ^{-1}. About 96% of the residues are available at cost below \$1.5 GJ^{-1}.

5.3.1.2 Crop Residue Costs Delivered at the Conversion Plant

Fig. 5.4 shows the combined cost supply curve for the corn and wheat residues at the factory gate delivered to the conversion plant for the various scenarios considered in the case study. These costs include crop residue harvesting, collection, baling, and storage at the farm, transport to a local distribution point as well as long distance transport by truck to the conversion plant. We included a base case scenario where bales are transported by truck from the farms to the conversion plant and improved scenarios where further preprocessing is undertaken close to the farms and rail transport is used for long-distance transport to the power plant.

On average, crop residues in South Africa are delivered at the power plant at a cost of about \$7.1 GJ^{-1}—this is a weighted average cost for

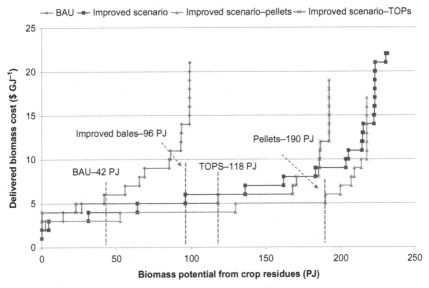

Figure 5.4 Cost–supply curve for corn and wheat residues delivered to the conversion plant in central South Africa. The dashed vertical lines indicate available crop residues below \$5 GJ^{-1} delivery costs.

biomass from all regions. About 11% of the biomass is delivered at the conversion plant at less than \$3 GJ^{-1}, whereas about 36% can be delivered at less than \$5 GJ^{-1}. About 82% is delivered at less than \$10 GJ^{-1} and only 5% of the biomass is delivered above \$15 GJ^{-1}.

At current conditions, the supply chain that delivers conventional pellets has the lowest cost biomass (\$4.1 GJ^{-1}) followed by TOPs (\$5.7 GJ^{-1}). As TOPs processing costs decline in the future, average delivered costs of TOPs are also expected to decrease to \$4.7 GJ^{-1}. The base case (with raw biomass bales) shows the highest delivered cost of \$6.9 GJ^{-1} compared to the improved case supply chain (\$6.6 GJ^{-1}). This is because train transport becomes more efficient with larger volumes of biomass (and longer distances traveled) associated with the improved case (raw bales).

Despite the additional preprocessing costs of biomass (\$13.3 t^{-1} for CPs and \$52.4 t^{-1} for TOPs), the pellet chain and TOPs chain deliver lower-cost biomass to the conversion plant as shown in Fig. 5.4. About 24% and 14% of pellets and raw bales, respectively, cost below \$3 GJ^{-1} at the factory gate. For TOPs, 12% is delivered at costs below \$3 GJ^{-1}. About 87% of CPs and TOPs are delivered at \$5 GJ^{-1}, compared to 42% of raw bales. About 92% of conventional pellet-based biomass is delivered at the factory gate at costs below \$6 GJ^{-1}, while corresponding values for TOPs and raw bales are 89% and 60%, respectively. Nearly all CP- and TOP-based biomass (99%) is delivered below \$10 GJ^{-1}.

In absolute terms, only 42 PJ is delivered in the base case scenario below \$5 GJ^{-1} compared to 96 PJ (for raw biomass-improved scenario), 190 PJ (CPs) and 168 PJ (TOPs). Therefore, considering cofiring 30% biomass at Camden (1600 MWe out) requires 36 PJ biomass feedstock—at current conditions, there is adequate biomass below \$5 GJ^{-1} to meet this demand. For this particular power plant, supplies can therefore be built up over time with changing demand and improvements in supply.

5.4 CONCLUSIONS

Global bioenergy markets have been growing and key bioenergy feedstocks such as wood pellets are becoming global commodities that are traded on the international markets. The EU is likely to remain the center for bioenergy markets, key importer and driver of solid biomass trade especially for the power and heat applications, given the projected growth in biopower and challenges to meet demand with local biomass production in the region. North America, eastern Europe, and Russia are already supplying woody biomass to the EU, and in the future, Brazil and coastal Africa are likely to be the major suppliers of biomass feedstock as well. Local markets in these producing regions will also become important as bioenergy technologies mature.

Trade will remain an important enabler for developing the bioenergy sector by facilitating the production and supply across regions. Optimized logistics based on an efficient transport system and preprocessed biomass feedstock are key to the delivery of competitive biomass. From the results of the case studies, in the short term, conventional pellets are still expected to play an important role as the internationally traded solid biomass commodity and can also in the longer term be cost-effectively used as a feedstock in biopower production. In the near future, torrefied pellets may become the dominant and preferred internationally traded solid biomass commodity as the technology is commercialized. This should result in improvements in bioenergy supply chains both in terms of costs and greenhouse gas impacts. Therefore, in the short term, it would be more cost-effective to ship densified solid biomass from different regions of the world where low-cost biomass is available to the major bioenergy markets for final large-scale conversion, given the advantages of economies of scale offered by these markets and risks of market immaturity for developing large-scale power production from biomass in the major biomass-producing regions. Local markets are expected to develop also, and provide localized opportunities for local biomass producers and conversion plants. Utilization of agricultural residues for power production is a promising option, but its application in cofiring configuration is yet to be proven on a wide commercial scale. More investigations are required to establish the technical feasibility and economics of large-scale mobilization of agricultural residues for such bioenergy applications.

Given the distribution of biomass production regions and markets as well as the nature of raw biomass, preprocessing biomass plays an important role in improving biomass supply chain economics. Logistics and transport are key cost components in the biomass value chain and major investments in infrastructure and capacity are required to realize large-scale biomass supplies. Establishing this infrastructure is gradual and takes time, which also applies to the mobilization of large volumes of biomass. These two aspects are interrelated and region-specific due to the unique settings for biomass feedstock production and local infrastructure. Given this context, there is a need for examining the entire biomass supply value chain so as to understand the many elements involved in bioenergy mobilization.

REFERENCES

Agar, D., Wihersaari, M., 2012. Bio-coal, torrefied lignocellulosic resources. Key properties for its use in co-firing with fossil coal-Their status. Biomass Bioenerg 44, 107–111.

Artec Biotechnologie GmbH, 2015. Available at: http://www.artec-biotechnologie.com/, Last visited: 11.06.2015.

AVA-CO2 Forschung GmbH, 2015. Available at: http://www.ava-co2.com/web/pages/de/downloads.php, Last visited: 11.06.2015.

Bagramov, G., 2010. Economy of Converting Wood to Biocoal. Lappeenranta University of Technology, Lappeenranta, Finland113., MSc Thesis.

Batidzirai, B., Mignot, A.P.R., Schakel, W., Junginger, M., Faaij, A.P.C., 2013. Biomass torrefaction technology—status and future prospects. Energy 62, 196–214.

Batidzirai, B., van der Hilst, F., Meerman, J.C., Junginger, M., Faaij, A.P.C., 2014. Optimisation of biomass supply chains with torrefaction technology. Biofuel. Bioprod. Bior. 8, 253–282. http://dx.doi.org/10.1002/bbb.1458.

Beekes, M., Cremers, M., September 2012. Realising a co-firing dream. Power Eng. Int. 20 (8), 64–70.

Ben, H., Ragauskas, A.J., 2012. Torrefaction of Loblolly pine. Green Chem 14, 72–76. (2012).

Bergman, P.C.A., 2005. Combined torrefaction and pelletisation: the TOP process. ECN-C--05-073.

Boyd, T., de Vries, D., Kempthorne, H., Wearing, J., Wolff, I., 2011. Mass & Energy Balance for Torrefied Pellet Production. UBC Biomass Pelletization Workshop, May 18, 2011.

Chum, H., Faaij, A., Moreira, J., Berndes, G., Dhamija, P., Dong, H., et al., 2011. Bioenergy (Chapter 2). In: Edenhofer, O., Pichs-Madruga, R., Sokona, Y., Seyboth, K., Matschoss, P. (Eds.), IPCC Special Report on Renewable Energy Sources and Climate Change Mitigation Cambridge Cambridge University Press, UK and New York, USA, pp. 209–332.

Clemens, A., Klemm, M., Backes, M., 2012. Hydrothermal carbonisation of organic waste. 4th International Conference on Engineering for Waste and Biomass Valorisation, Porto 10.-13. September 2012.

Cocchi, M., Nikolaisen, L., Junginger, H.M., Goh, C.S., Heinimö, J., Bradley, D., et al., 2012. Global Wood Pellet Industry Market and Trade Study. Accessed on 03/04/2015. Available at: http://www.bioenergytrade.org/downloads/t40-global-wood-pellet-market-study_final_R.pdf.

Crocker, M., Andrews, R., 2010. The rationale for biofuels. In: Crocker, M. (Ed.), Thermochemical Conversion of Biomass to Liquid Fuels and Chemicals The Royal Society of Chemistry, Cambridge, UK, pp. 1–25.

CS carbonSolutions Deutschland GmbH, 2015. Available at: http://www.cs-carbonsolutions.de/, Last visited: 11.06.2015.

DBFZ. 2013. Internal analytic report.

Deng, J., Wang, G.-J., Kuang, J.-H., Zhang, Y.-L., Luo, Y.-H., 2009. Pretreatment of agricultural residues for co-gasification via torrefaction. J. Anal. Appl. Pyrol. 86 (2), 331–337.

Dinjus, E., Kruse, A., Tröger, N., 2011. Hydrothermale Karbonisierung: 1. Einfluss des Lignins in Lignicellulosen. Chemie Ingenieur Technik 10 (2011).

DOE—Department of Energy (South Africa), 2011. Integrated Resource Plan for Electricity 2010-2030. Revision 2. Final Report (25 March 2011).

Dornburg, V., Faaij, A., Verweij, P., Langeveld, H., van de Ven, G., Meeusen, M., et al., 2010. Bioenergy revisited: key factors in global potentials of bioenergy. Energ Environ Sci 3, 258–267. 2010.

EC (European Commission), 2014a. Impact Assessment. Communication from the Commission to the European Parliament, the Council, the European Economic and

Social Committee and the Committee of the Regions. A policy framework for climate and energy in the period 2020 up to 2030. SWD(2014)15, Brussels, 22.1.2014. Available at: http://eur-lex.europa.eu/legal-content/EN/TXT/PDF/?uri=CELEX:52 014SC0015&from=EN.

EC (European Commission), 2014b. State of play on the sustainability of solid and gaseous biomass used for electricity, heating and cooling in the EU. SWD(2014) 259 final. Brussels, 28.7.2014.

Escala, M., Zumbühl, T., Koller, Ch, Junge, R., Krebs, R., 2013. Hydrothermal carbonization as an energy-efficient alternative to established drying technologies for sewage sludge: a feasibility study on a laboratory scale. Energy & Fuels 27 (1), 454–460.

EverGreen Renewable, LLC, August 2009. Biomass Torrefaction as a Preprocessing Step for Thermal Conversion—Reducing Costs in the Biomass Supply Chain.

Europe's Energy Portal, 2013. Fuel Prices, Rates for Power & Natural Gas. Available at: www.energy.eu/. Accessed 06//20/ 2013.

Faaij, A., Domac, J., 2006. Emerging international bio-energy markets and opportunities for socio-economic development. Energy for Sustainable Development 10, 7–19.

Faaij, A., Junginger, M., Goh, C.S., 2014. Chapter 1: A General Introduction to International Bioenergy Trade. In: Junginger, M., Goh, C.S., Faaij, A. (Eds.), International Bioenergy Trade, Springer.

Funke, A., and Ziegler, F., 2010. Hydrothermal carbonization of biomass: a summary and discussion of chemical mechanisms for process engineering—Biofuels, Bioprod, Bioref 4:160–177.

Goh, C.S., Cocchi, M., Junginger, M., Marchal, D., Daniela, T., Hennig, C., et al., 2012. Wood pellet market and trade: a global perspective. Biofuel, Bioprod. Bioref 7 (1), 24–42.

Gunarathne, D.S., Mueller, A., Fleck, S., Kolb, T., Chmielewski, J.K., Yang, W., et al., 2014. Gasification characteristics of hydrothermal carbonized biomass in an Up-draft Pilot-Scale Gasifier. Energy Fuels 28 (3), 1992–2002. http://dx.doi.org/10.1021/ef402342e.

Hamelinck, C.N., Suurs, R.A.A., Faaij, A.P.C., 2005. International bioenergy transport costs and energy balance. Biomass Bioenerg 29 (2), 114–134.

Hitzl, M., Corma, A., Pomares, F., Renz, M., 2014. The Hydrothermal Carbonization (HTC) plant as a decentral biorefinery for wet biomass. Catalysis Today 11/2014. http://dx.doi.org/10.1016/j.cattod.2014.09.024.

IEA, 2010. Projected Costs of Generating Electricity—2010 Edition. International Energy Agency/Nuclear Energy Agency/Organisation for Economic Co-Operation and Development. Paris. pp. 218(2010).

IRENA, 2012. Financial Mechanisms and Investment Frameworks for Renewables in Developing Countries, December 2012. Available at: http://irena.org/Finance_RE_Developing_Countries.pdf. Last viewed 20/06/2014.

Junginger, M., Faaij, A., van den Broek, R., Koopmans, A., Hulscher, W., 2001. Fuel supply strategies for large-scale bio-energy projects in developing countries. Electricity generation from agricultural and forest residues in Northeastern Thailand. Biomass Bioenergy 21 (4), 259–275.

Kiel, J., Zwart, R., and Verhoeff, F., 2012. Torrefaction by ECN. Presentation to the SECTOR/IEA Bioenergy Torrefaction Workshop, 20th European Biomass Conference and Exhibition, June 21, 2012, Milan, Italy.

Kietzmann, F., Klemm, M., Blümel, R., Clemens, A., 2013. Demonstrationsanlage für HTC-Kohle in Halle—abfallwirtschaftliche Ziele und technische Umsetzung. In: Wiemer, K., Kern, M., Raussen, T. (Eds.), Bio- und Sekundärrohstoffverwertung VIII stofflich—energetisch. Neues aus Forschung und Praxis Witzenhausen-Institut, Witzenhausen, pp. 337–347.

Klemm, M., Clemens, A., Kietzmann, F., Blümel, R., Nelles, M., 2015. Hydrothermale Verfahren—sinnvolle Ergänzung oder Irrweg? 14. Münsteraner Abfallwirtschaftstage Münsteraner Schriften zur Abfallwirtschaft Bd, pp 16.

Kline, K.L., Oladosu, G.A., Wolfe, A.K., Perlack, R.D., Dale, V.H., McMahon, M., 2008. Biofuel Feedstock Assessment for Selected Countries. ORNL/TM-2007/224, Oak Ridge National Laboratory/Appalachian State University/US Department of Energy, pp 135 + Appendices.

Kretschmer, B., Allen, B., Hart, K., 2012. Mobilising Cereal Straw in the EU to Feed Advanced Biofuel Production. May 2012, Report produced for Novozymes. IEEP, London.

Lamers, P., Junginger, H.M., Hamelinck, C., Faaij, A., 2012. Developments in international solid biofuel trade—an analysis of volumes, policies, and market factors. Renew Sust Energ Rev 16 (5), 3176–3199.

Lamers, P., Marchal, D., Heinimö, J., Steierer, F., 2014. Chapter 3: Global Woody Biomass Trade for Energy. In: Junginger, J., Goh, C.S., Faaij, A. (Eds.), International Bioenergy Trade, Springer.

Lamers, P., Hoefnagels, R., Junginger, M., Hamelinck, C., Faaij, A., 2015. Global solid biomass trade for energy by 2020: an assessment of potential import streams and supply costs to North-West Europe under different sustainability constraints. GCB Bioenergy 7 (4), 618–634.

Li, J., Brzdekiewicz, A., Yang, W., Blasiak, W., 2012. Co-firing based on biomass torrefaction in a pulverized coal boiler with aim of 100% fuel switching. Appl Energy 99, 344–354.

Libra, J.A., Kern, J., and Emmerich, K.H., 2011. Hydrothermal carbonization of biomass residuals: a comparative review of the chemistry, processes an application of wet and dry pyrolysis—Biofuels, pp. 89–124.

Luo, X., 2011. Torrefaction of biomass—a comparative and kinetic study of thermal decomposition for Norway spruce stump, poplar and fuel tree chips. Swedish University of Agricultural Sciences, Uppsala, MSc Thesis.

Matzenberger, J., Kranzl, L., Tromborg, E., Junginger, M., Daioglou, V., Sheng Goh, C., et al., 2015. Future perspectives of international bioenergy trade (Review). Renew Sust Energ Rev 43, 926–941.

Meerman, J.C., Ramírez, A., Turkenburg, W.C., Faaij, A.P.C., 2012. Performance of simulated flexible integrated gasification polygeneration facilities, Part B: economic evaluation. Renew Sust Energ Rev 16 (8), 6083–6102.

Melin, S., 2011. Torrefied Wood—A New Emerging Energy Carrier. Presentation to Clean Coal Power Coalition CCPC March 9, 2011.

Miao, Z., Shastri, Y., Grift, T.E., Hansen, A.C., Ting, K.C., 2012. Lignocellulosic biomass feedstock transportation alternatives, logistics, equipment configurations, and modelling. Biofuel Bioprod Bioref 6, 351–362.

OECD/IEA, 2010. IEA Energy Technology perspectives 2010. International Energy Agency, Paris. 458.

OECD/IEA, 2011. World Energy Outlook 2011. IEA, Paris. 666.

Peidong, Z., Yanli, Y., Yongsheng, T., Xutong, Y., Yongkai, Z., Yonghong, Z., et al., 2009. Bioenergy industries development in China: dilemma and solution. Renew Sust Energ Rev 13, 2571–2579.

Perlack, R.D., Wright, L.L., Turnhollow, A.F., Graham, R.L., Stokes, B.J., Erbach, D.C., 2005. Biomass as Feedstock for a Bioenergy and Bioproducts Industry: The Technical Feasibility of a Billion-Ton Annual Supply. DOE/GO-102995-2135, ORNL/TM-2005/66. US Department of Energy/US Department of Agriculture/Oak Ridge National Laboratory, Oak Ridge, pp 47 + Appendices.

Phanphanich, M., Mani, S., 2011. Impact of torrefaction on the grindability and fuel characteristics of forest biomass. Bioresource Technol 102 (2), 1246–1253.

Purohit, P., 2009. Economic potential of biomass gasification projects under clean development mechanism in India. J Cleaner Prod 17, 181–193.

Ramke, H.-G., Blöhse, D., Lehmann, H.-J., Antonietti, M., Fettig, J., 2010. Machbarkeitsstudie zur Energiegewinnung aus organischen Siedlungsabfällen durch Hydrothermale Carbonisierung. Final report for Deutsche Bundesstiftung Umwelt, Höxter.

REN21, 2011. Renewables 2011 Global Status Report. Paris: REN21 Secretariat.

Rentizelas, A., Tolis, A., Tatsiopoulos, I.P., 2009. Logistics issues of biomass: the storage problem and the multi-biomass supply chain. Renew Sust Energ Rev 13 (4), 887–894.

Schipfer, F., Bienert, K., Kranzl, L., Majer, S., Nebel, E., 2015. Deployment scenarios and socio-economic assessment of torrefied biomass chains. Part 2: Results. Deliverable, SECTOR project. Available at: https://sector-project.eu/fileadmin/downloads/deliverables/SECTOR_D9.5_final.pdf. Last viewed 14/05/2014.

Sevilla, M., 2009. The production of carbon materials by hydrothermal carbonization of cellulose, A.B. Fuertes Carbon, pp. 2281–2289.

Sikkema, R., Steiner, M., Junginger, M., Hiegl, W., Hansen, M.T., Faaij, A.P.C., 2011. The European wood pellet markets: current status and prospects for 2020. Biofuel Bioprod Bior 5 (3), 250–278.

Skøtt, T., 2011. Straw to Energy—Status, Technologies and Innovation in Denmark 2011. August 2011. Accessed on 03/04/2015. Available at: http://inbiom.dk/Files//Files/Publikationer/halmpjeceuk_2011_web.pdf.

Smeets, E.M.W., Faaij, A.P.C., Lewandowski, I.M., Turkenburg, W.C., 2007. A bottom up quickscan and review of global bio-energy potentials to 2050. Progress in Energy and Combustion Science 2007 (33), 56–-106.

Statistisches Bundesamt, 2013. Abfallbilanz, Available at: <https://www.destatis.de/DE/ZahlenFakten/Gesamtwirtschaft   Umwelt/Umwelt/UmweltstatistischeErhebungen/Abfallwirtschaft/Tabellen/Abfallbilanz2010.pdf?__blob=publicationFile>. Last visited: 05.24.2013.

SunCoal Industries GmbH, 2015. Available at: http://www.suncoal.de/, Last visited: 11.06.2015.

Tarcon, R., 2011. Canadian perspective on biomass production and distribution. IEA Cofiring Workshop, January 25–26, 2011, Drax Power Station, UK. Accessed on 15/03/2013. Available at: http://www.iea-coal.org.uk/publishor/system/component_view.asp?LogDocId=82487&PhyDocId=7694.

Tapasvi, D., Khalil, R., Skreiberg, Ø., Tran, K., Grønli, M., 2012. Torrefaction of Norwegian birch and spruce: an experimental study using Macro-TGA. Energ Fuels 26, 5232–5240.

TerraNova Energy GmbH, 2015. Available at: http://www.terranova-energy.com/index.php, Last visited: 11.06.2015.

Trading Economics, 2014. Average interest rates for Mozambique. Accessed on 12/03/2014. Available at: http://www.tradingeconomics.com/mozambique/interest-rate.

Tremel, A., Stemann, J., Herrmann, M., Erlach, B., Spliethoff, H., 2012. Entrained flow gasification of bio-coal from hydrothermal carbonization. Fuel 102, 396–-403.

Tumuluru, J.S., Hess, J.R., Boardman, R.D., Wright, C.T., Westover, T.L., 2012. Formulation, pretreatment, and densification options to improve biomass specifications for co-firing high percentages with coal. Industrial Biotechnology 8 (3), 113–132.

Tustin, J., 2012. IEA Bioenergy Annual Report 2011IEA Bioenergy:ExCo:2012:01. Available at: http://www.ieabcc.nl/publications/IEA%20Bioenergy%202011%20Annual%20Report.pdf. Accessed on 12/03/2014.

Tyndall, J.C., Berg, E.J., Colletti, J.P., 2011. Corn stover as a biofuel feedstock in Iowa's bioeconomy: an Iowa farmer survey. Biomass Bioenerg 35, 1485–1495.

Urošević, D.M., Gvozdenac-Urošević, B.D., 2012. Comprehensive analysis of a straw-fired power plant in the province of Vojvodina. Thermal Science 16 (1), S97–S106.

USDOE, 2012. Biomass: Multi-Year Program Plan. US Department of Energy, Office of Energy Efficiency and Renewable Energy. Available at: http://www1.eere.energy.gov/biomass/key_publications.html#PLANS_ROADMAPS_AND_REPORTS. Last visited: 12.03.2014.

Uslu, A., Faaij, A.P.C., Bergman, P.C.A., 2008. Pre-treatment technologies, and their effect on international bioenergy supply chain logistics. Techno-economic evaluation of torrefaction, fast pyrolysis and pelletisation. Energy 33 (8), 11206–11223.

van den Broek, M.,Veenendaal, P., Koutstaal, P., Turkenburg, W., Faaij, A.P.C., 2011. Impact of international climate policies on CO2 capture and storage deployment. Illustrated in the Dutch energy system. Energ Policy 39, 2000–2019.

Verhoest, C., and Ryckmans, Y., 2014. Industrial Wood pellets Report. Pellecert Project. Intelligent Energy Europe. Accessed on 24/03/2013. Available at: http://www.enplus-pellets.eu/wp-content/uploads/2012/04/Industrial-pellets-report_PellCert_2012_secured.pdf

WBGU (German Advisory Council on Global Change), 2009. World in Transition—Future Bioenergy and Sustainable Land Use. Earthscan, London.

Wilèn, C., PerttuJ., Järvinen T., Sipilä K.,Verhoeff F., Kiel J., 2013. Wood torrefaction—pilot tests and utilisation prospects. VTT Technology 122. pp. 73, Espoo (ISBN 978-951-38-8046-0).

CHAPTER 6

Commodity-Scale Biomass Trade and Integration with Other Supply Chains

E. Searcy[1], P. Lamers[1], M. Deutmeyer[2], T. Ranta[3], B. Hektor[4], J. Heinimö[5], E. Trømborg[6] and M. Wild[7]

[1]Idaho National Laboratory, Idaho Falls, ID, United States
[2]Green Carbon Group, Hamburg, Germany
[3]University of Lappeenranta, Lappeenranta, Finland
[4]Svebio, Stockholm, Sweden
[5]Mikkeli Development Miksei Ltd, Mikkeli, Finland
[6]Norwegian University of Life Sciences, Ås, Akershus, Norway
[7]Wild & Partner, Vienna, Austria

Contents

Abstract

A global bioeconomy requires adequate logistical infrastructure to support trade of biomass feedstock and intermediates. An integration of biomass trade streams with existing supply chain infrastructure, originally constructed for other goods, presents an opportunity to efficiently enable such growth. This chapter examines to what extent existing logistical infrastructure can be used and/or shared with biomass trade streams via specific case studies. It identifies how biomass trade is already or could be integrated into existing supply chains handling infrastructure, and for what kind of biomass specifications a dedicated infrastructure is needed. It finds that the existing solids handling infrastructure is well suited to integrate biomass intermediates such as conventional or torrefied pellets. Liquids with a higher energy density than solids, for example, pyrolysis oil, could potentially realize many opportunities to leverage

Developing the Global Bioeconomy.
DOI: http://dx.doi.org/10.1016/B978-0-12-805165-8.00006-9

© 2016 Elsevier Inc.
All rights reserved.

infrastructure designed for the petroleum industry, and may even enable leveraging home heating infrastructure, for example, in the US northeast, preventing costly modifications. However, high oxygen levels render pyrolysis oil corrosive, requiring investments in stainless steel or other more durable handling equipment. Biomethane injection into natural gas grids is already a common technology in most of Europe, but major hurdles remain, including high production costs, pipeline access, and the lack of quality standards.

6.1 INTRODUCTION

An increase in international biomass trade requires adequate logistical infrastructure that can support the respective quantities. An integration of biomass trade streams with existing supply chain infrastructure constructed for other goods presents an opportunity to efficiently enable such growth; up to a point where absolute quantities might justify dedicated biomass trade infrastructure. This chapter examines to what extent existing logistical infrastructure can be used and/or shared with biomass trade streams. Through specific case studies, it identifies how biomass trade is already or could be integrated into existing supply chains, and for what kind of biomass specifications a dedicated infrastructure might make more sense.

6.2 EVOLUTION OF COMMODITIZED BIOMASS

Most of the energy produced in industrialized countries comes from fossil energy commodities such as coal, oil, and natural gas. Over time, these supply systems have been standardized and commoditized, capturing substantial increases in liquidity and reduced transaction costs (Vactor, 2004). Biomass, on the other hand, in addition to low energy density, shares attributes with agricultural commodities, like vulnerability to weather and climate, and the need for timing of delivery to prevent spoilage or energy-value loss. Commoditization occurs as producers and intermediaries realize benefits from market integration and product standardization, and commodity exchanges develop systems to measure and grade goods.

A commodity is interchangeable with other products of the same type and most often used as an input to the production of other goods or services. For example, grain (a commodity) is used to make bread and other higher-value products before being sold to the end-user. The quality of a given commodity may differ slightly or may have variations, but it is essentially uniform across producers. When they are traded on an exchange, commodities must also meet specified minimum standards, also known as a basis grade.

As biomass production and trade volumes have increased and bio-energy industries have begun to evolve, minimum product standards for ethanol, plant oils, biodiesel, wood pellets, and energy wood chips have emerged (see, eg, Table 6.1). These specifications, although evolving as

Table 6.1 Most important standards for solid biomass

Standard code	Standard content
BS EN 14774-1:2009	Solid biofuels—Determination of moisture content—Oven dry method. Total moisture: Reference method
BS EN 14774-2:2009	Solid biofuels—Determination of moisture content—Oven dry method. Total moisture: Simplified method
BS EN 14774-3:2009	Solid biofuels—Determination of moisture content—Oven dry method. Moisture in general analysis sample
BS EN 14775:2009	Solid biofuels—Determination of ash content
BS EN 14918:2009	Solid biofuels—Determination of calorific value
BS EN 14961-1:2010	Solid biofuels—Fuel specifications and classes—Part 1: General requirements
BS EN 15103:2009	Solid biofuels—Determination of bulk density
BS EN 15148:2009	Solid biofuels—Determination of the content of volatile matter
BS EN 15210-1:2009	Solid biofuels—Determination of mechanical durability of pellets and briquettes. Pellets
CEN/TS 14588:2004	Solid biofuels—Terminology, definitions, and descriptions
CEN/TS 14778-1:2005	Solid biofuels—Sampling—Part 1: Methods for sampling
CEN/TS 14778-2:2005	Solid biofuels—Sampling—Part 2: Methods for sampling particulate material transported in lorries
CEN/TS 14779:2005	Solid biofuels—Sampling—Methods for preparing sampling plans and sampling certificates
CEN/TS 14780:2005	Solid biofuels—Methods for sample preparation
CEN/TS 15104:2005	Solid biofuels—Determination of total content of carbon, hydrogen, and nitrogen—Instrumental methods

(Continued)

Table 6.1 (Continued)

Standard code	Standard content
CEN/TS 15105:2005	Solid biofuels—Methods for determination of the water soluble content of chloride, sodium, and potassium
CEN/TS 15149-1:2006	Solid biofuels—Methods for the determination of particle size distribution—Part 1: Oscillating screen method using sieve apertures of 3.15 mm and above
CEN/TS 15149-2:2006	Solid biofuels—Methods for the determination of particle size distribution—Part 2: Vibrating screen method using sieve apertures of 3.15 mm and below
CEN/TS 15149-3:2006	Solid biofuels—Methods for the determination of particle size distribution—Part 3: Rotary screen method
CEN/TS 15150:2005	Solid biofuels—Methods for the determination of particle density
CEN/TS 15210-2:2005	Solid biofuels—Determination of mechanical durability of pellets and briquettes—Part 2: Briquettes
CEN/TS 15234:2006	Solid biofuels—Fuel quality assurance
CEN/TS 15289:2006	Solid biofuels—Determination of total content of sulphur and chlorine
CEN/TS 15290:2006	Solid biofuels—Determination of major elements
CEN/TS 15296:2006	Solid biofuels—Calculation of analyses to different bases
CEN/TS 15297:2006	Solid biofuels—Determination of minor elements
CEN/TS 15370-1:2006	Solid biofuels—Method for the determination of ash melting behavior—Part 1: Characteristic temperatures method

our understanding of feedstocks and conversion technologies grows, have allowed traders to place orders with different and interchangeable producers. Based on such standards international trade of various biomass fuels has increased steadily over recent years (Fig. 6.1).

There is a wide range of biomass potentially suitable for energy use (Chum et al., 2011; DOE, 2011), however many conversion systems have a narrow range of feedstocks that are compatible with their conversion process (eg, Davis et al., 2013; Jones et al., 2013). Quality standards are vital to enable reliable, efficient, trouble-free operation by end-users.

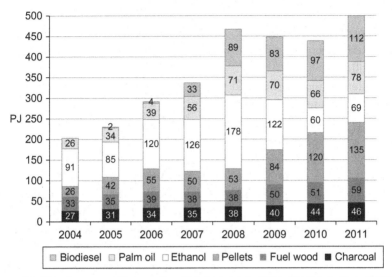

Figure 6.1 Development of international biofuels trade 2004–11 (Heinimö et al., 2013).

Even within a specific type of biomass, such as wood chips, there can be major differences in characteristics and properties between different batches chipped using different chippers, from different material, with different moisture content (Kenney et al., 2013). This variability may result in one batch using a particular piece of equipment to operate according to specification, but another may cause blockages in the fuel feed line, inefficient operation, harmful emissions, condensation in the flue, or automatic shutdown of the equipment as it moves outside its designed operating regime. In different equipment, however, the second batch of fuel may be perfectly acceptable. In addition to pure technical standards, sustainability standards are also required, and gradually being developed to ensure that biomass is only derived from sustainable, legal sources (see chapter: Sustainability Considerations for the Future Bioeconomy for details). Systems handling liquids and solids require fundamentally different equipment.

6.3 CURRENT COMMODITY-SCALE BIOMASS TRADE

With increasing use of biomass for energy, particularly when mandated via green electricity or transport fuel quotas, sourcing strategies (ie, the minimization of procurement costs) of large-scale users such as energy utilities or refineries have expanded into international markets. For some, this internationalization was relatively seamless as their value chains and operations were already global, as, for example, some biofuel producers that originated in the agricultural sector (eg, Archer Daniels Midland, Bunge, or Cargill).

At present, there is no solid biomass trade stream for energy that could classify as a commodity as trade is organized almost exclusively via contractual relationships, including (supra-)regional trade flows of fuelwood, wood chips, waste/residual wood, and wood pellets. Even international, cross-continental wood pellet trade, for example, between North America and Europe, is done via long-term contracts. Also, previous Rotterdam-based (spot-market) trading platform for wood pellets was unsuccessful (see chapter: Commoditization of Biomass Markets for a detailed assessment).

Furthermore, while the different wood pellet markets, that is, residential heating (over 15 million t in 2014), power and combined heat and power production (7.5 and 2 million t respectively in 2014) (EPC, 2015), are becoming increasingly intertwined, they are still organized separately and have different characteristics and (technical and sustainability) standards. To create a commodity market, additional integration would be required (see chapter: Commoditization of Biomass Markets for a detailed assessment). The residential heating market is predominantly supplied regionally via barge, rail, and truck transport. Large-scale bulk shipments, particularly overseas shipments, are almost exclusively used for direct delivery to deep sea ports (eg, Rotterdam) where they are rerouted for delivery to power stations.

The European Union (EU) is the largest wood pellet market, accounting for almost three-quarters of total global consumption. Its local production of 13.5 million t (of 27.1 million t total production in 2014) is fully consumed internally plus an additional 5.3 million t of extra-EU imports in 2014 (EPC, 2015). Intra-EU trade is dominated by exports from Latvia, Lithuania, Romania, Spain, and Portugal. International shipments of wood pellets currently originate mainly in North America, destined for use in Europe and Southeast Asia. The United States and Canada exported 5.15 million t of wood pellets to Europe in 2014 (EPC, 2015). With the introduction of new regulations in South Korea, future exports to this region are uncertain.[1]

Trade of commodity-type feedstock intermediates is currently limited to the liquid biofuel sector, that is, ethanol and fatty-acid methyl

---

[1] http://biomassmagazine.com/blog/article/2015/09/south-korean-market-on-ice (accessed October 1, 2015).

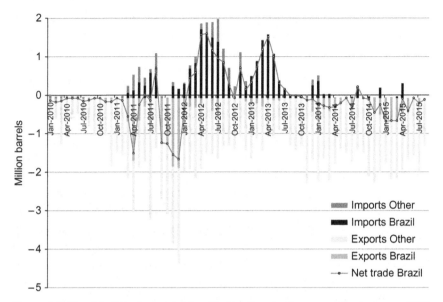

Figure 6.2 Monthly US ethanol trade with Brazil and other countries between 2011 and 2013. *Data from EIA, 2015. Petroleum and Other Liquids—Imports and Export Statistics. Washington, DC: US Energy Information Administration (EIA).*

ester, biodiesel, or respective vegetable oil trade (Lamers et al., 2014). The United States and Brazil are the two largest producers and exporters of ethanol in the world, with ethanol being produced from corn feedstocks in the United States and sugarcane in Brazil.[2] By 2010, increasing corn harvests and limited US ethanol market growth led the United States to become a net exporter of ethanol and the world's main supplier (Fig. 6.2). At the same time, in Brazil, decreased sugarcane harvests slumped local ethanol production and reversed the traditional global ethanol trade pattern (ie, a domination of Brazilian exports). Brazil required US ethanol imports to meet domestic demand in 2011. Brazilian production recovered in 2012 and exports to the United Stated grew again, largely due to growing US Renewable Fuel Standard (RFS2) targets, under which sugarcane fuel ethanol is considered an advanced biofuel.

Global bioenergy markets and trade flows continue to grow and intertwine. Still, neither ethanol nor biodiesel (or wood pellets) are even close

[2] For production statistics see: http://www.ethanolrfa.org/resources/industry/statistics/ (accessed October 1, 2015).

to the level of market maturity of fossil energy commodities like hard coal or mineral oil. In essence, there is still plenty of room for further commoditization of biomass into more feedstocks with a variety of specifications compatible with different conversion processes.

6.4 THE INTEGRATION OF COMMODITIZED BIOMASS WITH OTHER COMMODITY SUPPLY CHAINS

There exists an extensive high-capacity handling and transportation infrastructure that has been developed and refined for other industries, designed to handle solids, liquids, or gases. This same infrastructure, potentially requiring small modifications, could be leveraged to reduce cost and secure feedstock supply in an expanding bioenergy industry, particularly for international trade. However, additional steps to format the material such that it can use the existing equipment are often required. This section discusses some of the challenges and opportunities associated with this infrastructure.

6.4.1 Leveraging Solids Handling Infrastructure

This section gives several examples of leveraging solids handling infrastructure—using infrastructure designed for grain to move and handle bulk solids, namely pellets, and using infrastructure designed to transport and handle logs to move woody residues. Both cases offer opportunities for cost savings, however both require an additional step or steps to format the material such that it can use the existing equipment.[3]

6.4.1.1 Case Study 1: Conventional and Torrefied Wood Pellets

As outlined above, an international pellet trade industry exists today. One advantage of pelletizing biomass, primarily wood, prior to export, is that pellets can be handled similarly to grain. Once the biomass is formatted into pellets, augers, storage units, and transport containers developed for grain can, with small modifications and handling adjustments, it can be used to efficiently move pellets over long distances. Pelletization also increases energy density and grindability and facilitates reactor in-feed.

---

[3] Portions of this section are based on Bradley, D., B. Hektor, M. Wild, M. Deutmeyer, P.-P. Schouwenberg, J.R. Hess, J.S. Tumuluru, and K. Bradburn (2014). Low Cost, Long Distance Biomass Supply Chains – Update April. IEA Bioenergy Task 40: Sustainable International Bioenergy Trade. C.S. Goh and M. Junginger.

Wood pellets have traditionally been made from compressed sawdust or other residuals from sawmilling and manufacturing other wood products, though an increasing proportion of raw material for pellets now comes from pulpwood and to a limited extent harvest residues. Pellets are manufactured in several types and grades as fuels for electric power plants, district heating, homes, and other applications (Melin, 2011). Pellets are dense and have to be produced with a low moisture content (below 10%) that allows them to be burned with a very high combustion efficiency (Cocchi et al., 2011). Pellets are produced in different sizes, normally between 6 mm and 8 mm in diameter (ENplus). The regular geometry and small size of pellets allow automatic feeding with very fine calibration using an auger feeding or by pneumatic conveying. Their high density also permits compact storage and economic transport over long distance. Pellets can be moved conveniently from a bulk vessel to a storage bunker or silo on a customer's premises (Melin, 2011).

Handling and transporting pellets has risks similar to those of other biotic material (grain, wood chips, lumber, etc.), primarily linked to off-gassing and dust explosion. Rules for safety and risk reduction have been established, for example, under the IMO (International Maritime Organization), and users continue to develop and apply new procedures to minimize risk, including safety initiatives by different industry associations. While some serious accidents have occurred in wood pellet transportation, it should be noted that many accidents have occurred for other energy forms as well, including coal, oil, and natural gas.

Torrefying biomass may offer additional benefits to leverage infrastructure beyond that of pelleting biomass. In particular, increased grindability (ie, improved brittleness) facilitates reactor in-feed at the biorefinery for thermochemical conversion processes, particularly when cofiring with coal. Increased energy density, easier handling through hydrophobic character, and increased water resistance, lower off-gassing, and reduced self-heating properties are expected to result in a significant reduction in transportation costs.

Torrefaction is a thermo-chemical process whereby biomass, such as wood, is heated under an inert or nitrogen atmosphere within a temperature range of 230–300°C, similar to roasting coffee beans. Improved ease of pulverization, increased water resistance, increased energy density, and cleaner burning are significant advantages of torrefation. Torrefied biomass is lightweight, causing plenty of fines and dust if transported, therefore mechanical compacting of torrefied biomass (such as by pelletization or briquetting) is necessary to make it transportable. All solid biomass can be torrefied.

Torrefied biomass is hydrophobic, however, torrefied biomass in pellet or briquette form will not necessarily be waterproof. The quality of the compacting process will determine how long the products can withstand the eventual entrance of water into cracks and rifts in the surface of the product. Water encroachment does not necessarily result in complete decomposition of a torrefied pellet as it would with wood pellets, but it still weakens the physical strength of torrefied products and its grindability in coal power plants.

Under the near-term target, customers for biomass upgrading through torrefaction are large-scale coal power plants, as torrefied wood is easily integrated into existing coal conversion plants and logistical systems. Wood chips and wood/agro pellets only allow limited cofiring ratios or complete conversion of the plant, often only up to 10% before physical limitations come in to play, such as gumming up coal grinders. Some instances report up to 30% depending on the grade of adaptation of the feeding system, coal mills, and boilers. Also, plants are designed for a certain volume of throughput, and if fuels that are fed in have a lower calorific value than coal, the power output is reduced and the plant is derated. In laboratory scale, torrefied biomass has proven that 100% firing regimes are possible with minimum adjustments to the coal power plant's combustion unit and at significantly reduced derating compared to wood pellets.

One company, New Biomass Energy in the United States, is capable of producing at the commercial level and is now exporting small volumes by ship. Torrefied pellets have a higher calorific value and energy density than wood pellets (Table 6.2). As an example of leveraging existing infrastructure and technologies, wood pellets commonly are shipped on 35,000-t Handymax ships, and on occasion 70,000-t Panamax ships. If and when torrefied pellets are traded long distances, they will be shipped in small volumes in some holds of small ships as pellets were initially. Coal, a major commodity, is shipped from modern terminals on Panamax and Capesize ships. Major coal exporters Australia, Indonesia, the United States, Columbia, and western Canada also have considerable available biomass. It is envisioned that torrefied pellets can be manufactured near coal terminals, and subsequently be loaded into holds adjacent to coal on large ships, thereby getting the cost advantage of large vessels.

Compacted torrefied biomass is currently considered to be a nonhazardous good. Once cooled no increases in temperature have been reported

Table 6.2 Properties of transportable biomass and competing fossil fuels

	Fresh wood	Wood pellets	Torrefied pellets	Pyrolysis oil	Coal	Heavy fuel oil
Moisture (%)	35–50	7–10	1–5	20–25	10–15	<0.5
Calorific value (GJ/t)	9–12	16–18	19–23	16–19	23–28	42.5
Bulk density (tonne/m³)	0.2–0.25	0.6–0.68	0.65–0.75	1.2	0.8–0.85	0.99
Energy density (GJ/m³)	2–3	9.6–12.2	12.4–17.3	19.2–22.8	18.4–23.8	42
Acidity (pH)				2–3		
Ash (% by wt)		0.4–2	0.4–2.5	<0.25	9.7–20.2	0.08

to date when stored or shipped in bulk carriers. Off-gassing is similar to or less than that seen from wood pellets. Torrefied product traded today is transported and stored under special licences, as there is insufficient experience with the material. Registration such as REACH (Registration, Evaluation, Authorization & restriction of Chemicals) and licensing is now being assessed.

In summary, transforming raw biomass into pellets or torrefied pellets required additional processing steps and hence costs, however these steps improve fuel properties and enable the use of existing solid commodity handling infrastructure such as grain or coal, and may facilitate in-plant handling, including reactor in-feed. There is a pellet industry that exists today, mostly wood pellets going to Europe. Another opportunity for leveraging solids handling equipment can be found in the field, where bulky, loose residues are collected and handled.

6.4.1.2 Case Study 2: Residue Bundling System Integration With Forest Industry in Finland

In the case of woody biomass, there are opportunities to increase solids handling before leaving the forest. As an example of this, the forestry industry has developed equipment for moving logs out of the forest and to a plant (for pulp wood or lumber, for example). Operators in Finland have demonstrated using this logging equipment to handle and transport bulky woody residues that were distributed throughout the forest after delimbing. In order to leverage this equipment, the residues were densified into bundles.

Woody residues are the limbs and tops of trees, generally distrusted throughout the forest when trees are delimbed at the stump or at the landing to make logs. Woody resides may also include small trees. Bundlers compact woody residues into a cylindrical log, bunching them using twine or another binder to contain the material. The bundle can then be moved in bulk using existing handling equipment, such as a forwarder, and are a potential technology that could enable the leveraging of existing equipment in the biomass industry.

In Finland, 7.5 million cubic meters (solid m³) of forest chips were used at 910 combustion plants in 2014, accounting for 54 PJ or 4% of the country's total energy use (Ylitalo, 2015) and the target for the end of 2020 has been set at 13.5 million cubic meters (TEE, 2010). In 2014, the main method for forest-chip supply was roadside chipping directly into trucks (57%), with less material going via regional terminals (29%) or directly to end-user crushing stations (14%) (Strandström, 2015). Terrain chipping has almost disappeared, as have the logging residue bundling method types operated in early 2000.

The bundling system for logging residues was first commercialized in western Finland around the year 2000. Due to the long transport distances, large supply area, and large annual harvesting volumes, the circumstances for introducing the novel bundling production technology were favorable on the west coast of Finland. Also, the integration of bundle production into the supply of industrial roundwood was straightforward, and the synergies were significant because the first large-scale user, combined heat and power (CHP) plant Alholmens Kraft, was located within the large pulp and paper and saw-mill integrate of the forest industry company UPM (Poikola et al., 2002). Another benefit was that all the machines in the supply system were able to operate independently of each other, making the system more efficient and reliable. The bundling system also stabilized the biomass, improving storage characteristics, since the compact bundles did not start decomposing immediately. This made it possible to store them for the peak demand periods during winter in energy production. Moreover, the bundling of logging residues preserved the fuel quality and reduced space requirements during storage compared to storage of loose logging residues in windrows (Laitila et al., 2013).

In the bundling method, the forest residues, or slash, left behind by a harvester, are collected and fed into the bundler machine built on the for-warder carrier. The bundler produces compact residual logs kept tight with a cord. Typically composite residue logs (CRL) were around 3 m in length and about 60–80 cm in diameter, which was optimal for truck logistics. After being bundled, the CRL material was forwarded from the forest to the roadside with standard forwarders. CRL material was stored temporarily in the forest or the bundles were transported directly to the power plant by truck. CRL were usually crushed at the power plant or at terminal inventory, which enabled the use of very effective and economical crushing or chipping method.

At the landing, the logging residue logs were stacked alongside conventional timber assortments and transported with the standard timber trucks to the terminal or end-use facility. The unloading of the logging residue logs took place at the end-use facility with similar equipment to that for unloading saw logs or pulpwood. In the most efficient cases, the logging residue logs were unloaded directly from the timber truck to the feeding table of the stationary crusher.

In 2004, a whole-tree bundler (Fixteri), capable of incorporating compaction into the cutting phase, was launched and later in 2009 the second version and in 2012 the latest version, Fixteri FX 15 (Fig. 6.3). In addition

Figure 6.3 Fixteri FX 15 whole tree bundling supply system (Image rights: Fixteri Oy).[4]

[4] For details, see: http://www.fixteri.fi (accessed on December 3, 2014).

to bundles with pulpwood dimensioned trees, separate energy wood bundles composed of undersized trees and undesirable tree species were produced. Except for placing bundles onto the feeding table, the bundling process is autonomous, enabling simultaneous cutting and accumulation of subsequent bunches. Per hectare, up to 40% more sellable material can be obtained from the forest, when the harvesting is carried out in an integrated manner.

Although bundling increases costs in the forest, cost savings can be achieved over the whole system when relatively long forest forwarding and road transportation distances are required. The development of this system is ongoing and a number of technical improvements could considerably increase the competitiveness. Local circumstances as work conditions, forest types, transport distances, and end-user's infrastructure dictate the suitability of this concept.

Residues are a low-density bioenergy source that may be distributed throughout the forest. The addition of an extra densification step in the forest, namely bundling, enables the use of conventional forestry equipment, and may offer cost savings when moving the material over longer distances.

6.4.1.3 Case Study 3: Shipping of Forest Biomass Over the Baltic Sea

The above solids handling examples focused on moving solid biomass over land, however opportunities may arise to use shipping infrastructure as well. Shipping densified bulk solid biomass, such as logs (and therefore bundles) and pellets, is a potentially economically viable method of moving the material over long distances via a waterway. This method of course requires access to a waterway, and offers lower per-mile cost than truck or rail, however it has higher loading and unloading costs. Note that sawdust and chips are far below the optimum weight cargo.

Shipping of woody biomass is a mature activity in the Baltic Sea region. Trade of biomass and other goods has had and still has a key role in the economic and cultural development of the region. The Baltic Sea region has an infrastructure that includes vessels, ports, and others, and also a variety of actors, institutions, networks, and experience.

Although sizes vary, the typical vessel for transport of woody biomass in the Baltic is of 5000 deadweight tonnes (DWT), which means that it can take about $7000\,m^3$ in the hulls under deck and about $500\,m^3$ solid round-wood on deck.[5] A single trip is estimated to take 48 hours, and

---

[5] Information presented is based on empirical data and calculations. The analysis deals with a typical case based on average data, which can deviate from individual contracts.

shipping costs are based on long-term contracts for reasonably large quantities (>50,000 m³). Spot prices fluctuate a lot and are normally considerably higher. Once loaded onto the carrier, sea transport is much cheaper per mile than land transport.

Two typical types of loads are common: one with chips in the hulls and with roundwood on deck, the other with roundwood both in hulls and on deck. For suitable bulk carriers, the second option would mean lower transport costs. However, much of the energy woody biomass would emanate from chipped saw-mill waste or from logging residues, which often is cheaper than roundwood delivered to the export harbor. An increasing trade in crushed demolition wood and MSW is taking place in the Baltic Sea region, however relevant typical data are not available.

The wide range of variations in the physical properties of the various types of biomass shipped over the Baltic Sea, and the business issues related to quality and prices, call for close cooperation between sellers, buyers, agents, shippers, etc. on aspects that would increase the system efficiency of biomass trade. For example, trade of green chips (higher moisture content) is time-critical as temperatures below freezing may cause agglutination. Green chips may also undergo degradation and self-heating related to biological activity.

Nevertheless the integration of the supply chains of biomass in the form of energy wood chips and timber around the Baltic Sea using bulk carriers has been practiced successfully for many years and is a perfect example of how existing supply chains can be used to incorporate and integrate a growing trade in energy biomass and by doing so even optimizing overall transportation costs.

There is a substantial shipping infrastructure, including in the Baltic Sea region, which could be used to transport biomass overseas. This method, although having a low cost-per-mile, requires long distances to be economic, and requires the proper format to realize cost benefits and for safety reasons.[6]

6.4.2 Leveraging Liquids Handling Infrastructure

Densifying solid biomass into a liquid offers the potential to leverage liquids handling infrastructure. Biomass can be converted into a number of

---

[6] Portions of this section are based on Bradley, D., B. Hektor, M. Wild, M. Deutmeyer, P.-P. Schouwenberg, J.R. Hess, J.S. Tumuluru and K. Bradburn (2014). Low Cost, Long Distance Biomass Supply Chains – Update April. IEA Bioenergy Task 40: Sustainable International Bioenergy Trade. C.S. Goh and M. Junginger.

liquids including vehicle fuels, chemicals, and intermediates, including alcohols and sugars, each having its advantages and disadvantages. One potential liquid densification technology is pyrolysis, which produces three products: a liquid (pyrolysis oil), a solid char, and gas. There are many variations for pyrolysis, such as catalyzed fast pyrolysis and uncatalyzed fast pyrolysis. Pyrolysis oil may offer an opportunity to leverage residential infrastructure, as well as petroleum infrastructure.

Pyrolysis oil is twice as energy-dense as wood pellets, and is a free-flowing liquid. The oil is made from plant material by a thermochemical process whereby biomass particles are rapidly heated in the absence of oxygen (typically to 500°C in less than 1 second), vaporized, and the vapors then quenched into the pyrolysis oil liquid, also known as bio-oil. Pyrolysis oil does not resemble a vegetable or petroleum oil, because it is composed of hundreds of different chemicals including acids and contains approximately 25% water. Pyrolysis oil is immiscible with conventional fossil-derived oils but can be readily mixed with water. Pyrolysis oil can be combusted directly in boilers and potentially in engines (slow- to medium-speed diesels) and gas turbines for heat and power, although the latter applications are unproven beyond short-term testing using "enhanced" pyrolysis oil. Char is very fine and has low bulk density, around $250-350 \, kg/m^3$. As it can be difficult to handle in powder form, pelletizing char is recommended if transported any great distance. Pelletized char can be added directly to the coal feed without limitation or replace coal for reduction purposes in the metal industry.

Pyrolysis oil has been produced commercially in small volumes for around 25 years. Most production has been traded by truck throughout North America with only test volumes transported by ship, and in containers rather than in bulk. Pyrolysis oil could potentially be shipped to conventional petroleum refineries as a feedstock for upgrading into these lucrative markets.

Two of the world's leading pyrolysis oil companies, Ensyn and Dynamotive, have successfully demonstrated processing of palm oil extraction residues into pyrolysis oil. In Malaysia and Indonesia, the palm oil supply chain is already mature; truck or rail to ports, efficient storage and handling, and sufficient volume to warrant Panamax ships. Pyrolysis oil plants could be built near a palm plantation using the ready feedstock and with minor modifications use the existing ground supply chain to port.

Pyrolysis oil can be stored, pumped, and transported like petroleum products. However, pyrolysis oil has a low pH of around 2–3. The acidic and corrosive nature of pyrolysis oil means that modifications are required for storage and transportation. If oxygen is not removed from the oil, storage vessels and piping should be stainless steel, PVC, Teflon, or similar corrosion-resistant materials. Segregation/settling of pyrolysis oil is not an issue for short-term transportation and storage, and transport vessels are not required to have mixing capability. Mixing capability in customer storage tanks is easily arranged with existing tanks. One possible near-term application of pyrolysis oil is as a replacement for home heating oil.

Petroleum-derived heating oil is commonly used for home heating in the northeastern United States, and approximately US$15,000 is required to convert a single home from heating oil to an in-home wood pellet furnace (Bradley et al., 2014). As such, even when the economics of using solid biomass versus more expensive petroleum heating oil for home heating is favorable, the "in-home" capital investment and amortization time required for switching to solid biomass fuels (eg, pellets) is too great for most homeowners. For home heating oil applications in the northeastern United States, pyrolysis could potentially allow biomass to become an alternative and/or blend stock for the home heating oil market (DOE, 2012). However, challenges exist for converting raw biomass into a heating oil compatible product, including biomass specifications and variability; emergent pyrolysis conversion technologies and processes to produce stable and compatible bio-oils; infrastructure compatibility issues associated with corrosion, combustion, and fouling; bio-oil/heating oil blending and compatibility issues; and finally market acceptance and regulatory compliance. Challenges of corrosion, viscosity, and stratification can be overcome by hardening the existing infrastructure with corrosion-resistant tanks, pump systems, and modified high-volume burners, but such bio-oil distribution and in-home modifications are estimated to be about the same as converting to solid biomass fuels. The oxygen can be removed from the oil, mitigating corrosive issues such that the equipment would not have to be modified, however the oil would be more expensive.

In the oil industry, crude oil is either pumped long distances through pipelines or shipped on Capesize ships to refineries, where products such as gasoline, aircraft fuel, fuel oil, and chemicals are made. Stabilized pyrolysis oil could be similarly upgraded and blended at a refinery, leveraging billions of dollars in refinery and distribution infrastructure, although the oil would have to meet the refinery specifications.

Pyrolysis oil offers the potential to leverage residential home heating infrastructure and petroleum refining and distribution infrastructure, although both routes would either require equipment modifications or for the oil to be deoxygenated.[7]

6.4.3 Leveraging Gas Handling Infrastructure

Biomass can be used to produce gas via several different routes, such as gasification. Gas can also be collected from anaerobic degradation, such as those occurring in landfills and anaerobic digesters. Gases produced from biomass usually require cleaning and/or upgrading prior to conversion into another product, and prior to transport in a pipeline.

Biomethane, defined as methane produced from biomass and having properties close to natural gas, can in principle be used for exactly the same applications as natural gas (assuming the final composition is in line with the different natural gas qualities on the market). Therefore, it can be used as a substitute for transport fuels, to produce combined heat and power (CHP), heat alone, or serve as feedstock for the chemical sector. It can be transported and stored in the facilities and infrastructure available for natural gas. Biomethane can be produced by upgrading biogas or as so-called bio-SNG from thermochemical conversion of lignocellulosic biomass or other forms of biomass.

The technical feasibility to produce biomethane from biogas on a large scale has been demonstrated over the last decade. Table 6.3 gives an overview of the biomethane production in selected IEA member countries. To inject biogas into the natural gas grid or to use it as a vehicle fuel, the raw biogas has to be upgraded and pressurized. Biogas upgrading includes increasing the energy density by separating carbon dioxide from methane. Furthermore, water, hydrogen sulfide, and other contaminants are removed, sometimes before the upgrading process, to avoid corrosion or other problems in downstream applications. It is difficult to specify the exact characteristics for an upgrading technology, since the design and operating conditions vary between the different manufacturers, sizes, and applications. The key quality criteria for the upgrading technologies are the energy demand and the methane loss during upgrading.

---

[7]This section is partly based on Thrän, D., E. Billig, T. Persson, M. Svensson, J. Daniel-Gromke, J. Ponitka, M. Seiffert, J. Baldwin, L. Kranzl, F. Schipfer, J. Matzenberger, N. Devriendt, M. Dumont, J. Dahl, and G. Bochmann (2014). Biomethane—status and factors affecting market development and trade. M. Junginger and D. Baxter, IEA Bioenergy Task 40 and Task 37.

Table 6.3 Biomethane development in selected IEA bioenergy member countries in 2012 (Thrän et al., 2014)

Country	Biogas plants[a]	Biogas upgrading plants (feed-in)	Upgrading capacity[b] (Nm³/h)	Gas filling stations[c]	Gas-driven vehicles[d]
Austria	421	10 (7)	2000	203	7065
Belgium	119	0	0	15	355
Brazil	16[g]	n.d.	n.d.	1790	1,719,198
Canada	~50[e]	2 (n.d.)	400	83	14,205
Denmark	137	1 (1)	180[f]	4	81
Finland	34	5 (2)	959	18	1300
France	256	3 (2)	540[f]	149	13,300
Germany	9066	120 (118)	72,000	904	95,162
Ireland	22	0	0	0	3
Italy	1264	1 (0)	540[f]	903	746,470
Luxembourg	31	3 (3)	894[f]	7	261
Norway	44	5 (n.d.)	n.d.	23	353
South Korea	57	5 (n.d.)	1200[d]	184	39,000
Sweden	187	53 (11)	16,800[f]	190	44,000
Switzerland	600	16 (16)	n.d.	136	11,500
The Netherlands	211	16 (16)	6540[f]	150	5201
UK	265	3 (3)	1260[f]	40	520
USA	~440	25 (n.d.)	n.d.	1035	112,000
Total	>13,000	260 (>=179)	>100,000	>5800	>2,800,000

n.d., no data.
[a]Including waste water treatment plants, no landfill plants included.
[b]Referring to biomethane.
[c]Total (public and private).
[d]Motorcar, public transport, truck; natural gas vehicles (NGV).
[e]Only biogas plants, no data for waste water treatment plants available.
[f]Assuming 60% CH4 in the raw biogas.
[g]No waste water treatment plants and no landfill plants included.

The production of biomethane via thermochemical conversion is still in the pilot and demonstration stage, with no commercial market penetration so far.

The small-scale production of biomethane at many different locations is a fairly recent phenomenon, and requires additional efforts to adapt the regional infrastructure and to find adopted transport modes outside the natural gas grid. Biomethane may also play a significant role in future power-to-gas concepts by combination of renewable methane from excess energy, for example, by providing the renewable carbon source (separated CO_2), so that hydrogen produced from excess electricity and the renewable carbon source can be converted to methane, thus the overall methane output can be increased.

Even if the technical and logistical requirements for biomethane production are in principle available today, clear criteria for the biomethane quality to be fed into the gas grid and the end-use application are necessary. Compared to conventional fuels, the level of standardization is sparse for gaseous fuels. The International Organization for Standardization (ISO) has issued a natural gas standard, ISO 13686:1998 "Natural gas—Quality designation" and a standard for compressed natural gas, "ISO 15403 Natural gas—Natural gas for use as a compressed fuel for vehicles." The normative part of both standards contains no levels or limits, but have informal parts included with information for suggested values for gas composition, that is, from national standards or guidelines from France, Germany, the United Kingdom, and the United States. The absence of quantitative limits reflects the prevalent view of the gas industry that no precise gas quality can be specified, given the wide range of compositions of the raw gas obtained from underground.

There are a range of national standards in Europe for the injection of upgraded and purified biogas to the natural gas grid, and work on the international standardization of biomethane has been on-going. The specific challenge is to define standards which are attractive for the different potential end-users (gas grid owner, automotive industry, etc.) to enter the new market. Intensive discussions primarily concern sulfur and silicon content. Currently, two different standards for grid injection and automotive specification are under development at the European level and might be passed by the end of 2015.

One key driver for the application of biomethane is the reduction of greenhouse gas emission (GHG) due to the substitution of fossil fuels. Key parts in the production of biomethane that contribute to these GHG emissions include biomass cultivation and different biogas upgrading technologies, however sustainability standards and criteria do not always apply to biomethane.

Due to the complex supply chain, there are different environmental, economic, and administrative hurdles for the market introduction of biomethane. Today there are a wide range of approaches, instruments, and certificates established, which can differ in technical demands on grid injection and end-use, sustainability demands, support schemes, and monitoring of the biomethane flows.

Where gas pipelines or microgrids do not exist, biomethane can be transported in compressed or liquefied stage in mobile storage units (eg, in Sweden, see Thrän et al., 2014). The high cost of pipelines requires a high throughput in order to be cost-effective, and relatively smaller-scale biomethane facilities could not justify the additional cost. However, leveraging existing natural gas pipeline infrastructure would require clear quality standards in order for the pipeline owners to consider allowing the biomethane in their lines. Although substantial gas handling and transport infrastructure exists, the lack of quality standards remains a major barrier.

6.5 FUTURE TRENDS, RECOMMENDATION, AND CONCLUSION

There is extensive handling and transportation infrastructure that can be leveraged to increase the competitiveness of the bioenergy industry, including equipment for bulk solids, liquids, and gas. The demand of consuming parties (and technologies) for a higher refined product, especially for solid biomass, creates a side effect of the processed material now being able to piggyback on existing transportation infrastructure. This aspect, in combination with the increased scale of transportation resulting from growing demands, can enable transportation cost reduction per energy unit and increase the economically viable harvest and collection radius for processing operations.

As illustrated by the examples presented herein, there is a cost to processing the solid biomass such that it is compatible with grain handling infrastructure (in the case of pellets) and logging equipment (in the case of bundles). However, the investment in format would have additional gains downstream, beyond just reduction in handling, transport, and storage cost. As the technology exists today to make pellets and bundles, although the latter is a more nascent industry, this pathway has a relatively low barrier to entry as compared to the other phases.

Liquids have a higher energy density than solids, and pyrolysis oil could potentially realize many opportunities to leverage infrastructure designed for the petroleum industry, and may even enable leveraging home heating infrastructure in the US northeast, preventing costly modifications. However, high oxygen levels render the oil corrosive, requiring stainless steel or other more durable handling equipment. The oil can be stabilized by removing oxygen, however that would result in additional costs.

The opportunity to leverage gas handling and transportation infrastructure is likely the farthest away. Major hurdles remain, including high gas production costs, high barrier to entry into pipelines, and the lack of quality standards.

Regardless of the densified bioenergy format, it is clear that specifications and standardization are needed. Specifications currently exist for certain forms of bioenergy, such as wood chips, industrial energy wood pellets, residential wood pellets, bioethanol, and biodiesel. This list is likely to expand with a growing international bioenergy industry.

REFERENCES

Bradley, D., B. Hektor, M. Wild, M. Deutmeyer, P.-P. Schouwenberg, R., et al. (2014). Low Cost, Long Distance Biomass Supply Chains—Update April. IEA Bioenergy Task 40: Sustainable International Bioenergy Trade. C. S. Goh and M. Junginger.

Chum, H., Faaij, A., Moreira, J., Berndes, G., Dhamija, P., Dong, H., et al., 2011. Bioenergy. In: Edenhofer, O., Pichs-Madruga, R., Sokona, Y. (Eds.), IPCC Special Report on Renewable Energy Sources and Climate Change Mitigation Cambridge University Press, Cambridge, UK and New York, USA.

Cocchi, M., L. Nikolaisen, M. Junginger, C. Goh, J. Heinimö, D., et al. (2011). Global wood pellet industry and market study, IEA Bioenergy Task 40.

Davis, R., Tao, L., Tan, E.C.D., Biddy, M.J., Beckham, G.T., Scarlata, C., et al., 2013. Process Design and Economics for the Conversion of Lignocellulosic Biomass to Hydrocarbons: Dilute-Acid and Enzymatic Deconstruction of Biomass to Sugars and Biological Conversion of Sugars to Hydrocarbons. National Renewable Energy Laboratory, Idaho National Laboratory, Harris Group Inc, Golden, CO, USA.

DOE, 2011. U.S. Billion-Ton Update: Biomass Supply for a Bioenergy and Bioproducts Industry. In: Perlack, R.D., Stokes, B.J. (Eds.), Oak Ridge National Laboratory U.S. Department of Energy, Oak Ridge, TN, pp. 227.

DOE, U., 2012. Technical Information Exchange on Pyrolysis Oil: Potential for a Renewable Heating Oil Substitution Fuel in New England. U.S. Department of Energy, Manchester, NH, USA.

EIA, 2015. Petroleum and Other Liquids—Imports and Export Statistics. US Energy Information Administration (EIA), Washington DC, USA.

EPC, A., 2015. 2015 Pellet Market Overview AEBIOM Statistical Report. European Biomass Association & European Pellet Council, Brussels, Belgium.

Heinimö, J., P. Lamers, and T. Ranta (2013). International trade of energy biomass—an overview of the past development. 21st European Biomass Conference. Copenhagen, Denmark: 2029-2033.

Jones, S., Meyer, P., Snowden-Swan, L., Padmaperuma, A., Tan, E., Dutta, A., et al., 2013. Process Design and Economics for the Conversion of Lignocellulosic Biomass to Hydrocarbon Fuels: Fast pyrolysis and hydrotreating bio-oil pathway. Pacific Northwest National Laboratory, National Renewable Energy Laboratory, Idaho National Laboratory.

Kenney, K.L., Smith, W.A., Gresham, G.L., Westover, T.L., 2013. Understanding biomass feedstock variability. Biofuels 4, 111–127.

Laitila, J., Kilponen, M., Nuutinen, Y., 2013. Productivity and cost-efficiency of bundling logging residues at roadside landing. Croatian. Journal of Forest Engineering 34 (2), 175–187.

Lamers, P., Rosillo-Calle, F., Pelkmans, L., Hamelinck, C., 2014. Developments in international liquid biofuel trade. In: Junginger, M., Goh, C.S., Faaij, A. (Eds.), International Bioenergy Trade: History, Status & Outlook on Securing Sustainable Bioenergy Supply, Demand and Markets Springer, Berlin, pp. 17–40.

Melin, S., 2011. Research on Off-Gassing and Self-Heating in Wood Pellets During Bulk Storage. Wood Pellet Association of Canada, Vancouver, BC, Canada.

Poikola, J., C. Backlund, A. Korpilahti, K. Hillebrand, and S. Rinne (2002). The prerequisites of the bundling method in large scale wood fuel procurement. Puuenergian teknologiaohjelman vuosikirja 2002, Joensuu, Finland, 18-19, VTT symposium.

Strandström, M., 2015. Metsähakkeen tuotantoketjut Suomessa vuonna 2014. Metsätehon tuloskalvosarja, Metsäteho 8, 24.

TEE, 2010. Suomen kansallinen toimintasuunnitelma uusiutuvista lähteistä peräisin olevan energian edistämisestä direktiivin 2009/28/EY mukaisesti. Helsinki, Finland, Työ-ja elinkeinoministeriö, 10.

Thrän, D., E. Billig, T. Persson, M. Svensson, J. Daniel-Gromke, J., et al. (2014). Biomethane—status and factors affecting market development and trade. M. Junginger and D. Baxter, IEA Bioenergy Task 40 and Task 37.

Vactor, S.A.V. (2004). Flipping the Switch: The Transformation of Energy Markets. Ph.D., University of Cambridge.

Ylitalo, E., 2015. Wood in Energy Generation 2014. LUKE, Natural Resources Institute Finland.

CHAPTER 7

Commoditization of Biomass Markets

O. Olsson[1], P. Lamers[2], F. Schipfer[3] and M. Wild[4]

[1]Stockholm Environment Institute, Stockholm, Sweden
[2]Idaho National Laboratory, Idaho Falls, ID, United States
[3]Vienna University of Technology, Vienna, Austria
[4]Wild & Partner, Vienna, Austria

Contents

Abstract

Commodities are intermediate goods available in standardized qualities that are traded on competitive and liquid international markets. In this chapter, we analyze the current status and trajectories in biomass markets to discern to what extent solid biomass fuels are becoming commoditized. We present five criteria that are key indicators in the process towards commoditization and market maturity. These indicators are then

Developing the Global Bioeconomy.
DOI: http://dx.doi.org/10.1016/B978-0-12-805165-8.00007-0

© 2016 Elsevier Inc.
All rights reserved.

used as a framework to understand biomass market developments, with particular focus on wood pellet markets, and identify current obstacles to market maturity. We continuously draw comparisons with developments in fossil fuel markets with the dramatic developments in crude oil markets from the late 1960s to the mid-1980s used as a key example. In both the crude oil example and in the wood pellet discussion, the successful establishment of a futures contract is seen as a litmus test of the commoditization process. We find several similarities between historical and current fossil fuel markets and wood pellet markets in the reliance of vertical integration as a risk management tool and in how rigid fuel quality standards are perceived as obstacles to market liquidity. However, biomass markets also have particular characteristics that are not present in fossil fuel markets, especially the need for sustainability and traceability in supply chains. These are essential features of biomass fuels since their attractiveness to a very high degree relies on their being superior to fossil fuels in terms of lifecycle environmental performance. However, they do make the process of commoditization more difficult. For future discussions on biomass market developments, the tension here must be addressed.

7.1 INTRODUCTION

7.1.1 From Bioenergy to Bioeconomy

The past decade has seen a notable growth of bioenergy consumption in the Organisation for Economic Cooperation and Development region in the transportation, heat, and power sectors. This is a result of two dominating drivers: high and volatile fossil fuel prices and policy measures aimed at reducing greenhouse gas emissions to curb the growing threat of anthropogenic climate change. Bioenergy is expected to continue to increase its share in the global and European energy system (Chum et al., 2011). In addition, there is growing interest from both researchers and policymakers in the growing potential of biomass to replace fossil raw materials in other sectors, including but not limited to plastics, pharmaceuticals, and textiles (European Commission, 2012; McCormick and Kautto, 2013).

Biorefineries will play a key role in the large-scale implementation of the bioeconomy, and it is envisioned that biorefineries will replace the current role of petroleum-based refineries in the fossil-based economy (Kamm and Kamm, 2004). Developing efficient systems for the supply of biomass—the raw material input to the processes making up the backbone of the future bioeconomy—will be a key challenge if this vision is to become reality.

The prospect of commoditization of biomass is closely related to these discussions. Searcy et al. (2014) suggest that for biomass to be competitive with fossil materials and fuels, the former have to adopt the physical and institutional market infrastructure of the latter. The eventual emergence of the biobased economy will be highly dependent on the availability of standardized input goods, playing similar roles to those in the current fossil-based economy.

7.1.2 Commoditization of Biomass Markets

The term "commoditization" is increasingly used in relation to biomass markets to describe the transition of biomass-based fuel markets from local and informal to international and commercial (Wynn, 2011; Johnson, 2014). In this chapter, we will discuss the process of commoditization and what is required for an international biomass commodity market to emerge. Comparisons covering both physical/material properties and institutional/market-related issues are drawn with established commodity markets, thereby facilitating a discussion of the prospects of successful biomass commoditization.

7.2 DEFINING "COMMODITIES"

The terms "commodity" and "commoditization" are quite often thrown around without any clear definition. For the following discussions herein, we will use the definition by Clark et al. (2007, p. 3) who define a commodity as an "… intermediate good with a standard quality, which can be traded on competitive and liquid […] international physical markets."

This includes the key features that are most characteristic of commodity markets and will thus provide a useful starting point. Two properties of the commodity itself can be distinguished from the definition: commodities are intermediate goods and they are available in standard qualities. There are also properties related to actual market structures, namely that markets are competitive, liquid, and international. In the following, we will take this definition apart in order to more fully analyze what it entails.

7.2.1 Properties of the Good Itself

Commodities tend to be intermediate goods in the sense that they are not used primarily by final consumers, but instead as input in different forms of industrial processes. It is worth noting that these processes can take a wide range of different forms, ranging from combustion for generation of heat and power (energy commodities) to more intricate treatments in, for example, a petrochemical complex. Different end-users have different requirements in terms of input material quality but, in general, heterogeneity in inputs entails a higher risk of disturbance to the process itself or to end-product quality. This is especially true in industrial processes other than pure combustion-based energy generation. For this reason, standardization of input raw material quality is arguably the most crucial characteristic of commodities. Commodities are *fungible*, that is, "any one sample of 'it' would be interchangeable with any other sample of 'it.'" (Kub, 2014, p. 21). Many naturally occurring materials tend to be quite heterogeneous at the point of origin, be it an oil well or a copper mine. There are also goods, especially of biological origin, that are vulnerable to quality changes or deterioration during transport or storage. For these reasons, some sort of preprocessing is often necessary to achieve fungibility. A typical example is that oranges became a globally traded commodity not by trading the raw fruits but instead trading frozen concentrated orange juice (Newman, 2014).

If there are significant differences in a commodity related to physical qualities, this should be easily controlled and each commodity category should accordingly be divided into different standard qualities. For example, the coal market is divided into two main segments: steam coal and coking/metallurgical coal. Steam coal is used for energy purposes, primarily generation of electricity in condensing power stations. Coking/metallurgical coal, on the other hand, is used as a reducing agent in blast furnaces to remove oxygen from ore and produce pure iron.[1] Similarly, distinct grades of wheat are delineated by density and minimum percentages of specific wheat qualities (Clark et al., 2007; USDA, 2013). In the coal market, the separation of the steam coal market segment and the coking coal market segment is floating and market-determined. Sometimes, high-quality steam coal can be used as coking coal and conversely, if the coking coal market is weak but there is strong demand for steam coal, some coking coal shipments might be sold as steam coal (Cameron, 1997; Schernikau, 2010).

It is important to note that quality in this context is restricted to physical product-oriented quality, that is, ash content, moisture content, etc. Process-oriented quality criteria (ie, *how* the commodity has been extracted, cultivated or processed) have rarely—if everbeen an issue in established commodity markets (cf, Lima and Silveira, 2014). This is something that we will come back to later when discussing if and how traceability and sustainability certification can be combined with commoditization.

---

[1]Although it should be emphasized that coke is also used as fuel to generate the heat required in the smelting process.

7.2.2 Market-Related Properties

Whereas the first part of the definition above ("intermediate good with a standard quality") concerns the nature of the good itself, the second part ("...which can be traded on competitive and liquid global international markets") is related to how the good itself is traded between market actors. First of all, it is worth noting that a crucial consequence of the fungibility of commodities is that if all shipments are identical in terms of quality, *price* will be the sole criterion that determines whether a purchase will be made or not. In turn, this means that in order for a producing firm to be successful, focus on low costs will be a key component of business strategies in commodity markets (Porter, 2008).

As for the first of the criteria in our definition, commodity markets are competitive if there are many buyers and sellers and no single market actor or group of actors can exert enough influence to affect overall market prices. Market prices should thus be set according to the standard (theoretical) economical process of continuous fluctuations in overall market supply and demand balances and ensuing adjustments towards market equilibria (Clark et al., 2007).

Having said this, it is highly doubtful that truly competitive markets will ever exist in the real world. In markets for fossil energy commodities, there are certainly plenty of signs of imperfect market competition.

- For a long time, natural gas markets were quite heavily regulated, dominated by very long-term bilateral agreements and without much competition. There has certainly been a process towards increasing liberalization in recent decades, but the high reliance on pipelines for transportation still present opportunities for exertion of market power. There are significant differences between different continents as a result of the incomplete globalization of natural gas markets, where North American are more competitive than European. For example, the dominating role of Gazprom in the supply of natural gas to continental Europe has been the source of large controversy related to EU energy security and has been the subject of antitrust investigations by the European Commission (Kanter, 2014).

- In oil markets, the impact on prices from production agreements within the Organization of Petroleum Exporting Countries (OPEC) is well known, although the actual market power of OPEC certainly has fluctuated significantly in the time since the organization was first formed in the 1960s (Fattouh and Mahadeva, 2013). This is further elaborated upon in chapter "Biorefineries: Industry Status and Economics."

• The level of competitiveness in the global coal market differs—as for gas—between geographical markets. In the portion of global markets that includes coal actually traded internationally, a few large mining/commodity trading firms (Rio Tinto, BHP Billiton, Glencore, Xstrata)—that are also active in mining—dominate the supply side. However, the number of firms competing in the supply segment of the global coal market is significantly higher than in individual regions. For example, in 2006, four producers controlled 60% of European coal supply (Schernikau, 2010).

Market liquidity is an indication of how easy it is to convert an asset into cash without major adjustments to the price of the asset in question.[2] In order to facilitate market liquidity, it must be simple to acquire a reasonably correct value of the asset, that is, markets need to be transparent. Market transparency in turn is dependent on search costs, that is, the costs of acquiring adequate information necessary to take part in a transaction. A consequence of the fungibility of commodities is that search costs tend to be low, as comparisons between different offers are easy to make. In other words, standardization and fungibility are important in facilitating market transparency and in turn market liquidity (Carruthers and Stinchcombe, 1999).

However, another more important precondition for market liquidity is that there is adequate short-term demand for and adequate supply of the good in question. If there is no demand for a good that you are producing, it can obviously not be converted into cash at all. Since 85% of global energy demand is supplied by fossil fuels, the absence of a market is not a short-term issue for coal, oil, and natural gas. However, demand and supply differ between countries and regions depending on infrastructure (eg, a natural gas network) and energy policies.

The final part of the definition above relates to the geographical extent of markets, that is, the degree of international market integration. In fossil fuel markets, the level of market integration varies between fuels. In the crude oil market, more than 60% of global production is traded internationally, and for all practical purposes it can be seen as fully globalized even in a formal way (Kim et al., 2007). The same is mainly true for coal markets, at least the portions of the coal market actually traded between countries (Zaklan et al., 2012). Natural gas markets, which hitherto have been very dependent on fixed infrastructure in the form of pipelines, have been rather regionalized with only about 29% international trade,

---

[2] Cash itself is the most liquid asset.

although the increased role of liquefied natural gas will likely lead to an increasingly global natural gas market (Siliverstovs et al., 2005).

7.2.3 Futures Contracts: A Sign of Advanced Commoditization

Another key feature of considered commodities is the availability of futures contracts based on the physical commodity in question. Futures contracts can be used for physical supply, but are primarily used for price risk management and hedging. Establishing successful trade in a futures contract is by no means a simple affair. In fact, most futures contracts do in fact not draw sufficient market attention to be useful in the long-term and are thus withdrawn. In order for a futures contract to be successful, the fulfillment of the above-listed commodity criteria is generally deemed a necessity (Brorsen and Fofana, 2001). Thus, it is reasonable to see the successful establishment of a futures contract as a litmus test for the progress of a commoditization process.

7.3 COMMODITIZATION EXAMPLE: THE CASE OF THE CRUDE OIL MARKET

For our purpose of understanding if and how biomass markets might develop more towards a market structure characterized by fungibility, liquidity, transparency, and globally integrated markets, it might be useful to analyze the commoditization by using a historical example. In the following, we will review the commoditization process as it took place in the oil industry from the early 1970s to the mid-1980s.

7.3.1 The Commoditization of the Crude Oil Market From 1973 to 1987

Although the current situation with crude oil as an established commodity with an active futures market may be perceived as the normal order of things, the oil industry was dominated by vertically integrated companies for most of its history up until the 1970s. This was reflected in rigid price schemes and only small volumes actually being traded in the open market. Most oil flowed from wellhead to consumer within one single company, primarily one of the so-called Seven Sisters[3] that dominated the industry up until the 1970s (Parra, 2009; Yergin, 2009).

[3]The Seven Sisters in 1960 were BP, Royal Dutch/Shell, Gulf Oil, SoCal, Texaco, Esso, and Socony. Gulf, SoCal, and Texaco were later merged into what is now Chevron and Esso and Socony are today parts of ExxonMobil.

The historical prevalence of vertical integration as a means of organizing the crude oil industry is in fact not surprising, at least not if one looks at what is suggested by transaction cost theory (Williamson, 1979). Crude oil has a high degree of product heterogeneity and requires very capital-intensive infrastructure for processing. These are factors that are likely to result in industry organization relying on tightly coupled supply chains with vertical integration as a logical strategy.

The dominance of the Seven Sisters in the oil industry began to be challenged in the 1960s as new petroleum-rich countries started to develop their resources. In several cases, these new countries actively chose to encourage so-called independent (ie, non-Seven Sisters) oil companies to participate in development of new fields (Parra, 2009). This meant that growing—albeit still relatively quite small—volumes of crude oil were being handled outside the vertically integrated majors.

However, real change in the oil industry came in the 1970s with the oil crises in 1973–74 and 1979–80. The first crisis came in the middle of an ongoing wave of nationalization of oil resources, and consisted of two main components: (1) an OPEC embargo against the US and the Netherlands for their support of Israel in the Yom-Kippur war of October 1973 and (2) a significant reduction in Saudi Arabian oil production. Together, these two events had a massive effect on world energy markets as well as on the global economy, and were instrumental in highlighting the strengthening grip of OPEC countries over resources and prices. The Seven Sisters were losing nominal control over upstream resources, thereby involuntarily divesting the first stage of the vertically integrated supply chain (Van Vactor, 2004). Nevertheless, basic industry structures still remained by and large the same, as the producing countries were still reliant on the multinational oil companies for bringing the oil to market, with only marginal amounts sold and distributed through other channels (Fattouh, 2011).

This all changed when the second oil crisis came in 1979 as the Iranian revolution broke out and production in Iran dropped by 90% in a matter of weeks. British Petroleum (BP) in particular was very heavily dependent on oil from Iran, and when production more or less stopped, BP had to work frantically to make up for this shortfall to be able to live up to its own downstream obligations. The only way to do this was to try to outbid the other multinationals over crude from other countries, thereby further opening up the oil market into a vastly more dynamic structure (Van Vactor, 2004).

At the same time, OPEC tried to maintain a homogeneous price structure within the cartel, an ambition that was obviously challenged when individual OPEC countries took advantage of BP's scramble for volumes to make up the Iranian shortfall. Long-term contracts were broken which meant that the companies on the receiving end of these contracts in turn had to find these volumes elsewhere at higher prices. All this meant opportunities for a host of traders and brokers to move in to facilitate the large number of spot deals that followed (Fattouh, 2011; Van Vactor, 2004).

The consequence of this was a very chaotic market situation, with vast price dispersion as actual OPEC prices could vary in the 15–45$ range in December 1979. This was at the same time as the cartel tried to uphold a jointly administrated pricing system where the official benchmark OPEC price was only supposed to change after negotiations at official all-OPEC meetings (Van Vactor, 2004).

During the early years of the 1980s, oil demand dropped significantly at the same time as new production was coming online from the North Sea and elsewhere. These new suppliers gained market access by offering their crude in the spot market at lower prices than OPEC and hence OPEC's market share dropped from 51% to 28% in the time period (1973–85) (Fattouh, 2011). Spot market deals were also vastly growing as a share of the total market and prices in long-term contracts were to an increasing extent indexed to spot market prices. The consequence was an even closer link between oil prices and actual market fluctuations in terms of supply and demand balances. Moreover, in the early 1980s, price reporting agencies such as Platt's and Petroleum Intelligence Weekly started collecting and reporting crude oil spot prices on a weekly basis (Van Vactor, 2004). Information technology had also developed to the point where very rapid dissemination of information to market actors was possible, which further increased market transparency (Rost, 2015).

A crucial step in the commoditization process was taken when the New York Mercantile Exchange (NYMEX) successfully launched a crude oil futures contract in 1983. This had been attempted several times previously by different exchanges, but all previous launches had failed. However, NYMEX had a few years earlier been successful in launching a heating oil contract in a time when the energy market was obviously quite tumultuous already and when a cold winter in the US northeast further increased the need for price risk management tools. This paved the way for the later launch of the NYMEX crude oil futures contract, thereby putting the final major piece in the puzzle that makes up today's crude oil market (Van Vactor, 2004).

7.3.2 Analysis

If we go through our list of commodity criteria from Section 7.2 and analyze the oil example from this perspective, the main issue preventing the process of commoditization before the 1970s was arguably the lack of market competitiveness. As for the other criteria, oil markets were already very much international in the 1960s, standards had been developed since the 1920s and oil was obviously an intermediate good used to produce everything from asphalt to plastics. In addition, as most transactions were within a single firm, market liquidity was by and large a nonissue. However, the oligopolistic market structure and the fact that most oil flowed within the organizational walls of one of the Seven Sisters presented an insurmountable obstacle to further market development. Thus, it was not until the vertically integrated structures were ruptured that the commoditization process truly picked up speed. When (1) the OPEC countries started to assert control over oil resources and then—in the wake of the Iranian revolution—(2) spot market activity increased drastically, the barriers to commoditization were by and large removed.

7.4 COMMODITIZATION OF BIOMASS MARKETS

In this section, the prospects for commoditization of biomass markets are discussed by reviewing the commodity criteria in regards to current (mid-2015) status of markets for biomass fuels, wood chips, and wood pellets in particular.

7.4.1 "Intermediate Goods": What Is the Commodity Used For?

The first and perhaps least controversial commodity criterion is the definition that commodities are intermediate goods. This is true for both wood chips and wood pellets, although the classification of the latter merits some qualification as it is used in both industrial and consumer markets.[4]

Worth noting about wood chips is that they are used both for energy purposes and as raw material in the production of pulp and paper

---

[4]Wood pellets for residential consumption have some distinct characteristics that are not found in wood pellets traded for large-scale heat and power generation. These characteristics include some differences in quality requirements, but also an entirely different marketing structure where seasonality, warehousing, retailers, and branding are important components. The interaction between the large-scale market and the residential pellet market is found to be an interesting case to study, but to our knowledge this has not been done thus far.

products. This is worth discussing a bit further, especially in light of a development towards a biobased economy, where more possible end-users for biomass are expected to emerge and grow in significance. Here, there are interesting similarities with the market for coal and its division into steam coal and coking coal. In biomass markets, wood chips and precursors like pulpwood-quality roundwood might occasionally be used for energy purposes in weak pulp markets or bullish energy markets (Olsson, 2012). In other words, current market conditions decide whether the commodity in question is used for energy purposes or as an industrial input.

With the role of policy measures as a key determinant of biomass market developments, the issue of whether a certain shipment is used for energy or for industrial purposes has however become a rather contentious issue (Brännlund et al., 2010). In particular, in discussions about the actual implementation of a biobased economy, the issue of *cascading* is often brought up as a founding principle, emphasizing the importance that biomass resources are used in a way that maximizes resource efficiency (eg, European Commission, 2012). In this framework, which is based on a hierarchy of different uses, energy should be the last and final stage of the biomass use chain, that is, wood should first be used to produce lumber for, for example, construction, which at the end of its lifespan can be used to produce, for example, particle board and then, when the piece of particle board is finally to be discarded it should be combusted with energy recovery. As a principle, there is much to be said for this idea. However, there are discussions on an EU level of implementing the cascading principle in legislation and as a part of long-term strategies towards the biobased economy (European Commission, 2014). Actual implementation of such a framework is bound to be problematic, at least if past experiences in similar ventures are taken into account.[5]

The introduction of policy measures that aim to steer streams of fuels and raw materials in one direction or the other are difficult to implement with success. Category leakages and lack of policy cohesiveness between countries are likely to create loopholes in legislation that would require continuous patchwork and consequently more of the political uncertainties that have hindered stability in bioenergy markets.

---

[5]There are examples of similar regulations that have been problematic in several ways, for example, in the natural gas industry (Castillo Castillo, 2012; Talus, 2014) and also the Swedish wood fuel market in the 1980s and 1990s (Vinterbäck and Hillring, 1995).

The complexities arising from policy ambitions prioritizing some uses of biomass over others should thus not be underestimated, especially when it comes to these policy ambitions interfering with market functions. The political component is arguably the most significant risk in biomass markets when it comes to presenting obstacles to market development and commoditization. However, it is also worth contemplating the prospect of further segmentation of biomass markets between the use of biomass for energy and the use of biomass as (industrial) production process input. This would have important consequences not least for our next topic: standardization and product homogeneity.

7.4.2 Fungibility, Homogeneity, and Standardization
7.4.2.1 Product Quality
As noted above, fungibility is central to the definition of a commodity. Apart from being a technical property, minimizing uncertainty for the consumer about the characteristics of the raw material used in an industrial facility or a power station boiler, fungibility is also a vital precondition for market transparency, in turn a key facilitator of market liquidity.

The relative lack of progress in the commoditization of coal markets is often attributed to the unsatisfactory product homogeneity with large variations in coal quality between different mines and across the different categories of coal from lignite to anthracite. If the commodity is not homogeneous it can by definition not be a commodity. Lack of homogeneity is omnipresent in biomass fuels regardless of origin, albeit to a varying degree. Unrefined biomass fuels such as wood chips tend to be very heterogeneous in nature, with key properties like moisture content, ash content, ash composition, and energy density all depending on a host of different factors related to the actual geographical origin of the biomass, the (tree) species, and so on. Pelleting of wood and other biomass was a preprocessing innovation that made possible, among other things, a higher level of homogenization, which in turn means, for example, that a higher degree of automatization could be achieved in all further steps of handling and consumption (even small-scale wood pellet boilers can be automatized).

However, market actors in general and consumers in particular have for quite some time seen insufficient regularity in pellet quality as an obstacle to further market development. Processes aiming at biomass standardization to establish distinct standard qualities have been ongoing since the 1990s. Starting with national or regional wood pellet standards, these have since widened in geographical scope with a European standard (EN 14961-2) published in 2010 and a global (ISO 17225-2) in 2014. Successful adoption of standards by the market should be a key stepping stone in the development towards a commodity market for wood pellets.

At the same time, the relative immaturity of biomass markets makes standardization a complicated process, as there is the prospect of technological developments changing conditions drastically. Specifically, the process of torrefaction holds great promise when it comes to enabling further homogenization of biomass. Torrefaction is a preprocessing step that can remove heterogeneity in raw material input significantly, with torrefied biomass being quite similar to coal in many physical properties.

In other words, fungibility from the market perspective might result from both institutional development (standardization) and technological developments enabling the production of a more homogeneous good. It should also be noted that there is another possible route towards the end-goal of reducing risks related to the physical quality of raw material or fuel. Instead of achieving an ever-more homogeneous input, an alternative could be to build processes or develop combustion equipment that has a high level of tolerance for variations in fuel quality. In fact, fuel flexibility has been a widespread strategy across the heating and power sector, for example, in Sweden. Here, most newly constructed combined heat and power boilers are constructed to use a rather wide variety of fuels from municipal solid waste to wood chips.

As for the importance of commoditization for future biomass markets in a biobased economy, there are potential lessons to be learned from the separation of the coal market into steam/energy coal and coking/metallurgic coal. In the coking coal market, homogeneity and quality consistency is a key feature and a necessity for industrial processes to work smoothly, whereas there is more space for quality variations in the steam coal market. One example of the latter is of course the cofiring of wood pellets with coal. This does not mean that steam coal consumers are unconcerned with quality properties, but they do have more opportunities for flexibility. For example, coal of different qualities can be mixed to achieve a desired average level for which a boiler is optimized (Schernikau, 2010). Boilers can also be retrofitted to allow for more flexibility when it comes to coal quality. An example of this is the installation of scrubbers in many European coal-fired power stations, which made it possible to use less expensive high-sulfur coal without breaching air pollution norms (eg, Cameron, 1997). It is however interesting to note that some coal traders and market analysts have a tendency to think that the quality criteria are allowed to play too large a role and thus be a larger obstacle than necessary to interchangeability between different coals. This is attributed to exaggerated concerns among power station engineers about the actual economic value of homogeneity in fuel input (Cameron, 1997; Walters, 2010). Interestingly, the same kinds of discussions are now ongoing in reference to wood pellet markets, where some have argued that excessively tight quality criteria stipulated by large wood pellet consumers are creating an obstacle to further market development (Maroo, 2012).

Thus, the drive for the commoditization of wood pellets may not be strong enough as long as the generation of heat and power remains the major point of final consumption. In other words, commoditization may be further pushed by industrial processes with tight quality requirements, for example, cellulosic biofuel production. Here, conversion efficiencies are directly related to the quality of the input material and drop significantly, for example, under higher moisture and/or ash content (Argo et al., 2013; Muth et al., 2014). This increases production costs, reduces market competitiveness, and general economic viability of the business operation.

7.4.2.2 Commoditization, Sustainability, and Traceability

As noted above, a key characteristic of commodities is that any one sample or shipment of, for example, Brent crude oil should be no different than any other sample or shipment. In other words, it does not matter where, when and how the specific commodity batch was produced as long as it fulfills the quality requirements for the grade in question. This means that buyers and sellers need not engage in time- and resource-consuming negotiations or relationship building for each of them to be certain that the actual deal will satisfy their respective aspirations. The fungibility characteristic thus significantly reduces transaction costs and acts as a great facilitator of market transparency and market liquidity.

However, concurrently with the development of biomass commoditization, there is a drive towards increased lifecycle traceability in order to ensure sustainability throughout the supply chain. Whether or not further commoditization of biomass is possible in this environment is an important question. The current fragmentation of criteria between not only different certification systems but between countries and continents as well could create an obstacle to the transition towards biomass commoditization (Rodriguez et al., 2011). Commoditization of raw materials for the bioeconomy will require not only harmonization and standardization of physical product quality but of process-related qualities, that is, supply chain sustainability and traceability, as well. There have been suggestions to include sustainability criteria as a component in futures contracts for biofuels (Mathews, 2008) which should be feasible, but traceability is another question.

There are however also problems with current setups of systems to ensure sustainability of bioenergy supply chains. Currently, complex mandatory sustainability criteria are implemented in both the EU and the US to ensure that bioenergy raw materials are produced in a sustainable manner. However, most if not all of these raw materials are being cultivated and processed for other purposes (food, feed, industrial inputs, etc.) without the rigorous clout of sustainability requirements surrounding the bioenergy sector. This situation with the existence of two completely different sets of standards leads to risks of leakages between categories (Pelkmans et al., 2014). Although this is already problematic, the situation is particularly untenable in the light of an eventual development towards a biobased economy with biomass-based materials replacing fossil materials not only for energy purposes but as industrial inputs in general.

7.4.3 Market Structure

A general characteristic of biomass markets is that the bulky nature of biomass and the ensuing low economic value per volume unit make transportation costs a large share of overall costs. This has a tendency to reduce the geographical extent of biomass markets.

The international trade in biomass for energy has thus far primarily been an issue of international trade in wood pellets for use in large-scale generation of heat and electricity, where pellets are often replacing coal. Industrial wood pellet markets seem to inherit much of the basic structure of the international coal trade, but with an even higher degree of market concentration. Although no recent assessment of the market concentration is publicly available, a handful of European utilities (Drax, EON, RWE, Dong, Vattenfall, GDF-Suez) make up the vast majority of globally consumed pellet volumes.

From a global perspective, the vast majority of wood pellet producers operate rather small production facilities. However, a gradually larger share of global pellet production is being concentrated in a smaller number of key production facilities with the emergence of a rapidly expanding wood pellet production sector in the southeastern United States. US pellet production plants are several times larger than the largest European facilities (excluding the *Vyborgskaya* plant in the Russian Federation).

Another key characteristic of the transatlantic trade in wood pellets is that several of the large pellet plants constructed in the US are actually owned by the very same European utilities that are end-consumers of the pellets to be produced. As was noted in the discussion on the oil industry in Section 7.3.1, vertical integration can often be a natural strategy for managing both price risks and supply risks, especially in immature markets with heterogeneous products.

In addition, the industrial wood pellet market can also be classified as oligopolistic, at least on the buyers' side. The high level of market concentration in wood pellet markets is evident from the effects of the 2012 fire in RWEs wood pellet-fired power station in Tilbury. The damage from the fire meant that Tilbury was out of operation for several months. The power station would have consumed pellet volumes amounting to what was at the time a significant share of the total global market, volumes that then were sold into the open market, leading to a significant drop in prices (Kinney, 2012). The fact that events at a single facility had clear effects on overall market prices is a clear indication of the level of market concentration.

7.4.4 Market Liquidity

The concept of market liquidity and its status in biomass markets is closely related to fungibility but also to market concentration. An increased level of market transparency would be expected as a result of increased levels of standardization in biomass markets, especially wood pellets. Market liquidity is to a large extent a function of adequate supply of and adequate demand for the product in question. This was not an issue in the vertically integrated oil industry of the 1960s[6] as reviewed in chapter "Biorefineries: Industry Status and Economics," but the special characteristics of biomass markets (the policy dependence in particular) mean that both sides of the market are uncertain. Wood pellet demand has certainly been growing for several decades and the heating market segment continues to do so

---

[6] However, in the very early stages of the oil market development in the late 19th century, the industry was characterized by an almost continuous disequilibrium situation. Shortage was followed by surplus supplies which repeatedly caused large numbers of companies to bankrupt. This was largely due to the fact that development of infrastructure for distribution was lagging behind the growth in supply as well as demand (Sampson, 1975; Yergin, 2009).

at a solid pace. However, demand growth from the power market is quite erratic and dependent on policy decisions relating to policy measures supporting renewable electricity.

This means that, depending on the current policy environment, demand may be very strong or very weak, leading to general uncertainty that has clear implications for market liquidity. It is difficult to know, for example, if an investment in the pellet supply chain will make economic sense. It might not be possible to sell the produced pellets at a price necessary to make the investment profitable. Of course, these are the kinds of decisions that all businesses face, but the reliance of biomass markets on policy measures and the lack of long-term signals in, for example, EU policy regarding biomass for energy means that uncertainties are unusually high. A similar dilemma faces a power station considering converting its boilers from coal to pellets. The conversion will likely require large volumes of pellets, but whether or not these will be available in the open market is very difficult to know given the large overall level of market uncertainty. In this context, prevailing strategies of off-take contracts and vertical integration make sense, reducing risks at both ends of the supply chain. However, the process of facilitating bilateral contracts can be very arduous and time-consuming with each new transaction requiring specific negotiations and agreements on a large variety of conditions (Maroo, 2012; cf, Rodriguez et al., 2011; cf, Joskow, 1987).

7.4.5 International Market Integration

As for the level of international market integration in biomass markets, this is a topic that has thus far only been analyzed to a rather limited extent in comparison to the plethora of analyses of fossil fuel markets. There is no doubt that there is plenty of international trade in both wood chips and wood pellets, and in the latter case there are also very large intercontinental trade flows, especially from North America to Europe. It has been estimated that almost half of total global wood pellet production was traded between countries in 2010 (Lamers et al., 2012), which means that in this regard wood pellet markets are more internationalized than both coal and natural gas markets.

However, market integration is not only a matter of traded volumes, but rather—and actually even more so—of actual spillover of market fluctuations from one region to another. Whether this is the case is commonly investigated by analyzing price series over time to determine the extent to which prices comove. This methodology has been widely implemented in markets for coal (Zaklan et al., 2012), oil (Kim et al., 2007), and natural gas (Siliverstovs et al., 2005). As for wood fuel markets, the level of international market integration seems to be quite low. Only wood pellet markets in neighboring countries (Sweden–Denmark and Austria–Germany–Italy) show signs of market integration according to recent studies (Olsson et al., 2011, 2012; Olsson and Hillring, 2014; Bürger, 2015).

7.4.6 Conclusions

In this chapter, commoditization of biomass markets has been analyzed by focusing on the development of wood pellet markets and with the commoditization of crude oil as a point of comparison. It is important to note that this comparison only holds partial value as a result of the current fragmentation of the wood pellet market into (1) an industrial market focused on large-scale production, transport, and consumption (mainly for electricity generation) and (2) a more geographically constrained market with small-scale heating as the main source of demand. The two are not perfectly delineated or defined and there is a certain extent of flows between them, but in terms of market organization there are clear differences that must be acknowledged.

As with crude oil, it can be argued that vertical integration is the natural form of organization of at least the industrial portion of the wood pellet market.[7] Both oil and wood pellets can vary substantially in quality—although wood pellets arguably less so than crude oil—and both fuels are reliant on capital-intensive infrastructure and economies of scale to make economic sense. To our knowledge there is no up-to-date comprehensive information on the extent of vertical integration in the industrial wood pellet market. However, it certainly seems to be an increasingly common strategy for European utilities to invest in overseas wood pellet production facilities to thereby have control over almost the entire supply chain,[8] with RWE and Drax as key examples (Voegele, 2014). There are also examples of vertical integration being utilized as a strategy from the supply side. In August 2015, pellet producer German Pellets announced the acquisition of a Belgian power station planned to undergo conversion from coal to pellets (Argus Media, 2015).

---

[7] Production aimed at the small-scale premium market has not been as driven by economies of scale as the industrial segment of the market, although this may change with convergence between the two. See Section 7.5.2.2 for a more elaborate discussion on this.

[8] The "almost" qualification is needed since the utilities do not tend to own forest land itself, but rather stop their integration ambitions at the wood pellet production plant.

The question is: what might shift market developments to a higher share of spot trading in wood pellet markets? In the oil market, a host of factors combined to change (what seemed to be) the "natural" organization of the industry to something completely different. The decoupling of the resource ownership from the vertically integrated supply chain was obviously a key factor, but the tumultuous market swings and violent price fluctuations from 1973 to 1987 are arguably as important. The latter certainly stimulated the interest in short-term spot deals that could be done to draw use of a temporary favorable market opportunity that might only last a short while until things turned around. The information technology development of the 1970s and 1980s also made it possible to quickly access relevant market information. Brokers and traders were drawn into the market to facilitate these deals and profit from arbitrage opportunities. Similarly, the demand from market actors for stability and reduction of price risk paved the way for the establishment of the NYMEX crude oil futures contract. In the final section of this chapter we will outline the major obstacles that need to be removed for biomass to complete the commoditization process.

7.5 BIOMASS COMMODITIZATION: THE WAY FORWARD

7.5.1 Futures Contract Failure: A Sign of Market Immaturity

In addition to the five criteria that we used for the basis of analysis herein, a key characteristic of most commodity markets is the existence of a futures market based on the physical market of the commodity in question (cf, Siqueira et al., 2008).

In November 2011, APX-Endex in cooperation with the Port of Rotterdam introduced the possibility of trading a wood pellets futures contract based on their price index. This was generally seen as a groundbreaking move in terms of it being the first example of financial trading in a solid biomass fuel (Port of Rotterdam, 2011). However, the contract did not receive a large amount of interest from market actors and little or no actual trading in the contract carried out (Walet, 2012). In fact, as of the fall 2012 not a single trade in the contract had been conducted (Maroo, 2012) and less than 2 years after its introduction, ICE Endex discontinued the contract (ICE Endex, 2013).

It should be mentioned that a somewhat unusual feature of the APX-Endex contract was that physical delivery was mandatory, whereas in other futures markets this is an option that is available but rarely used. This is likely to have contributed to the failure of the contract (Rost, 2015).

At the same time, CME offers a wood pellet swap contract which in contrast is a strictly financial product with exchange only of financial flows and not of any physical pellets (Murray, 2012). Allegedly, trade in this contract has also been very sparse. Although it is important to note that most new futures contracts are terminated within 10 years of their introduction (Brorsen and Fofana, 2001), the reasons for the failed attempts at financial trade in wood pellets are likely to be found in current pellet market structures. The key question is: which are remaining obstacles and what steps need to be taken for these to be overcome?

7.5.2 The Road to Commoditization

Porter (2008) argues that "...products have a tendency to become *more like commodities* over time as buyers become more sophisticated and purchasing trends tend to be based on better information" (Porter, 2008, p. 3107). Does this mean that commoditization of biomass is only a matter of time? Or are there specific characteristics to biomass markets that will continue to present obstacles to commoditization over the foreseeable future?

7.5.2.1 Commoditization: Good for the Market but Not for All Market Actors?

It is important to note that for a single market actor, commoditization is by no means necessarily a desired development. If a good is fungible and competition is based on price only, this makes for a very competitive market, which is good for market function as a whole but not for all individual market actors. Instead, companies will often attempt to make the case that their specific brand is unique in one way or another and thus worth paying a premium for. As Roeber (1993, p. 77) put it "... transparency is the enemy of trading margins, and most companies would prefer to be partly-sighted in a foggy world."

Although Porter suggests that most goods over time tend to become commodities, this is not necessarily a one-way street. Sustainability as a concept is currently commonly being used in many product markets as a way to distinguish otherwise fungible commodities and thereby command a price premium (Lima and Silveira, 2014).

The control of prices and theoretically also quality was observed in the vertical integration of many value chains, including the previous examples of the oil and wood pellet industry. A similar pattern has been observed in the biofuel industry. For example, in Argentina, the key biodiesel producers have either ties to the upstream (vegetable oil production) or downstream part of the process (petroleum refineries).

7.5.2.2 Policy-Related Obstacles to Biomass Commoditization

For biomass to fulfill its potential not only as an energy source but also the raw material in an emerging bioeconomy, it is imperative to have long-term policies that work in support of this goal, not against it. In both the US and the EU, bioenergy policies have, in the recent decade, been erratic and constantly under revision. This creates an atmosphere of investment uncertainty that will hinder a sustained market development. The establishment of stable long-term policies that will work in a bioeconomy with biomass utilized in a wide variety of end-uses from food, to fodder, biochemical, and energy will be an exceptional challenge.

To make the bioeconomy vision and the prospect of biomass as a true commodity a reality, sustainability rules and regulations should be the same for all forms of biomass, regardless of their end-use. The seemingly micromanaging approach imposed, for example, by the current EU Renewable Energy Directive 2009/28/EC, is counterproductive. Technological development and the possibility to base the bioeconomy on a wide range of feedstock will make the distinctions between "biofuels," "bioliquids," and "solid biomass used for electricity, heating, and cooling" artificial and redundant. Introducing more specific rules to account for the variation in different end-uses and raw materials will only increase the probability of loopholes being exploited by savvy market actors. Similar problems are bound to arise from attempts of policy-induced rationing and earmarking of biomass resources for certain end-uses and not for others. Therefore, it is likely that the bioeconomy will require a complete reworking of the policy frameworks surrounding the use and trade of biomass resources. As this would obviously interfere with the established rigidities of agricultural tariff systems, it is however very difficult at this point to see how this would be carried out in practice.

7.5.2.3 Market-Related Obstacles to Biomass Commoditization

In the shorter term, it is worth contemplating the barriers to commoditization that contributed to the failure of the Rotterdam wood pellet futures contract. A key problem here is that the global wood pellet market is very much dominated by a handful of European utilities relying heavily on off-take contracts and vertical integration for management of both supply risk and price risk. Expanding the number of market actors, reducing the dominance of European utilities in the international wood pellet market, and thereby making the market more dynamic is necessary for continuing the commoditization process. A few pathways on how this might take place can be perceived.

One possibility would be the increased wood pellet demand from power producers in other continents. Here the recent years' growth of the Asian market could potentially inject more competition into the global pellet market. However, it is quite possible that the Asian market would develop into a separate market, separated from the European market by long distances and excessive transport costs. Shipping costs have also, in the last decade, proven to be very volatile, which is by itself a formidable obstacle to market integration and development. Innovations in logistics and/or preprocessing (eg, torrefaction) will likely be necessary to make possible sustained integration of the European and Asian wood pellet markets.

Another alternative would be the integration of the large-scale market of wood pellets for power production with the small-scale residential heating market. If wood pellet producers primarily focused on the large-scale market (1) are capable of producing pellets of a quality necessary for residential boilers and (2) have the proper channels to make their pellets available to residential consumers, this could strengthen the links between the two market segments. This could be an important way forward in terms of overall market development. The residential heating market is inherently characterized by a greater level of seasonal uncertainty as year-to-year demand can vary significantly depending on winter temperatures. In addition, the severity of winters can also vary between, for example, different parts of Europe, which also opens up arbitrage opportunities with trade being used to balance supply and demand. These uncertainties also open up for traders and brokers to play a more important role in providing market balance. Furthermore, a futures contract focused on pellets used in the European heating market was launched by Euronext in October 2015 (Euronext, 2015). It is quite possible that this approach could be more fruitful than the previous attempts at establishing futures trading based on the large-scale market. The number of individual transactions is obviously a lot higher in the small-scale heating market and the seasonality uncertainty increases demand for price risk management and hedging. Recall also from Section 7.3.1 that it was the successful establishment of a heating oil futures contract that paved the way for futures trading in crude oil. Similarly, it is quite possible that it is the small-scale heating market that will lead the way in terms of establishing more sophisticated means of price risk management for wood pellets.

ACKNOWLEDGMENT

Olle Olsson's contribution to this chapter has been supported by the Norden Top-level Research Initiative subprogram "Effect Studies and Adaptation to Climate Change" through the Nordic Centre of Excellence for Strategic Adaptation Research (NORD-STAR).

REFERENCES

Argo, A.M., Tan, E.C., Inman, D., Langholtz, M.H., Eaton, L.M., Jacobson, J.J., et al., 2013. Investigation of biochemical biorefinery sizing and environmental sustainability impacts for conventional bale system and advanced uniform biomass logistics designs. Biofuel. Bioprod. Bior. 7, 282–302. http://dx.doi.org/10.1002/bbb.1391.

Argus Media, 2015. Eon to sell Belgian plant to German Pellets. Argus Biomass Markets.

Brännlund, R., Lundmark, R., Söderholm, P., 2010. Kampen om skogen: koka, såga, bränna eller bevara? SNS förlag, Stockholm.

Brorsen, B.W., Fofana, N.F., 2001. Success and failure of agricultural futures contracts. J. Agribusiness 19 (2), 129–145.

Bürger, J., 2015. Preisstabilität oder -volatilität von forstlichen und industriellen Biomasse-Sortimenten am Beispiel ausgewählter Märkte MA-thesis, University of Life Sciences Vienna, Austria.

Cameron, J., 1997. International Coal Trade: The Evolution of a Global Market. OECD/IEA, Paris.

Carruthers, B.G., Stinchcombe, A.L., 1999. The social structure of liquidity: flexibility, markets, and states. Theory. Soc. 28, 353–382. http://dx.doi.org/10.1023/A:1006903103304.

Castillo Castillo, A., 2012. UKERC Technology and Policy Assessment Cost Methodologies Project: CCGT Case Study (No. REF UKERC/WP/TPA/2013/008). UK Energy Research Center, London.

Chum, H., Faaij, A., Moreira, J., Berndes, G., Dhamija, P., Dong, H., et al., 2011. Bioenergy IPCC Special Report on Renewable Energy Sources and Climate Change Mitigation. Cambridge University Press, Cambridge, UK and NewYork, NY, USA, pp. 209–332.

Clark, E., Lesourd, J.-B., Thiéblemont, R., 2007. International Commodity Trading: Physical and Derivative Markets, first ed. Wiley, Chichester.

Euronext, 2015. Euronext Launches Wood Pellet Contract. Biomass Magazine, Grand Forks, ND, USA.

European Commission, 2012. Innovating for Sustainable Growth: A Bioeconomy for Europe (Communication from the commission to the European parliament, the council, the European economic and social committee and the committee of the regions no. COM(2012) 60).

European Commission, 2014. Paper of the Services of DG Competition Containing Draft Guidelines on Environmental and Energy Aid for 2014–2020. European Commission, Brussels, Belgium.

Fattouh, B., 2011. An Anatomy of the Crude Oil Pricing System. Oxford Institute for Energy Studies, Oxford, UK.

Fattouh, B., Mahadeva, L., 2013. OPEC: What Difference Has It Made? (No. MEP 3). Oxford Institute for Energy Studies, Oxford.

ICE Endex, 2013. Industrial Wood Pellets Contract to Be Discontinued. ICE Endex Website.

Johnson, F.X., 2014. Exploiting Cross-Level Linkages to Steer the Bioenergy Transition (Doctoral Thesis). Royal Institute of Technology, Stockholm.

Joskow, P.L., 1987. Price Adjustment in Long Term Contracts: The Case of Coal. Dept. of Economics, Massachusetts Institute of Technology, Cambridge, Mass.

Kamm, B., Kamm, M., 2004. Principles of biorefineries. Appl. Microbiol. Biotechnol. 64, 137–145. http://dx.doi.org/10.1007/s00253-003-1537-7.

Kanter, J., 2014. Stalled Gazprom Antitrust Case May Suggest Unease for Energy Sanctions. NY. Times. (Print).

Kim, J., Oh, S., Heo, E., 2007. A Study on the Regionalization of the World Crude Oil Markets Using the Asymmetric Error Correction Model. Presented at the European Conference of the International Association of Energy Economists, Florence.

Kinney, S.-A., 2012. The Fallout From RWE's Tilbury Biomass Power Plant Fire. Forest2Market.   <http://www.forest2market.com/blog/The-Fallout-from-RWEs-Tilbury-Biomass-Power-Plant-Fire>.

Kub, E., 2014. Mastering the Grain Markets: How Profits Are Really Made. Kub Asset Advisory, Inc., Omaha, Nebraska.

Lamers, P., Junginger, M., Hamelinck, C., Faaij, A., 2012. Developments in international solid biofuel trade—an analysis of volumes, policies, and market factors. Renew. Sustain. Energ. Rev. 16, 3176–3199. http://dx.doi.org/10.1016/j.rser.2012.02.027.

Lima, A.C., da Silveira, J.A.G., 2014. Fairtrade and de-commoditization. In: Enke, M., Geigenmüller, A., Leischnig, A. (Eds.), Commodity Marketing Springer Fachmedien, Wiesbaden, pp. 449–463.

Maroo, J., 2012. Lack of critical mass in biomass Energ. Risk. <http://www.risk.net/energy-risk/feature/2213358/lack-of-critical-mass-in-biomass>.

Mathews, J.A., 2008. Towards a sustainably certifiable futures contract for biofuels. Energ. Policy 36, 1577–1583. http://dx.doi.org/10.1016/j.enpol.2008.01.024.

McCormick, K., Kautto, N., 2013. The bioeconomy in Europe: an overview. Sustainability 5, 2589–2608. http://dx.doi.org/10.3390/su5062589.

Murray, G., 2012. Low-risk trading environment Can. Biomass. <http://www.canadian-biomassmagazine.ca/news/low-risk-trading-environment-3571>.

Muth, D.J., Langholtz, M.H., Tan, E.C.D., Jacobson, J.J., Schwab, A., Wu, M.M., et al., 2014. Investigation of thermochemical biorefinery sizing and environmental sustainability impacts for conventional supply system and distributed pre-processing supply system designs. Biofuel. Bioprod. Bior. 8, 545–567. http://dx.doi.org/10.1002/bbb.1483.

Newman, K., 2014. The Secret Financial Life of Food: From Commodities Markets to Supermarkets. Columbia University Press, New York, NY.

Olsson, O., 2012. Wood fuel Markets in Northern Europe [WWW Document]. Available from: <http://pub.epsilon.slu.se/8859/> (accessed 24.09.14.).

Olsson, O., Hillring, B., 2014. The wood fuel market in Denmark: price development, market efficiency and internationalization. Energ. Int. J. 78, 141–148.

Olsson, O., Hillring, B., Vinterbäck, J., 2011. European wood pellet market integration—a study of the residential sector. Biomass Bioenerg. 35, 153–160. http://dx.doi.org/10.1016/j.biombioe.2010.08.020.

Olsson, O., Hillring, B., Vinterbäck, J., 2012. Estonian-Swedish wood fuel trade and market integration: a co-integration approach. Int. J. Energ. Sector Manag. 6, 75–90. http://dx.doi.org/10.1108/17506221211216553.

Parra, F., 2009. Oil Politics: A Modern History of Petroleum. I. B. Tauris, London; New York.

Pelkmans, L., Goovaerts, L., Goh, C.S., Junginger, M., Dam, J., van, Stupak, I., et al., 2014. The role of sustainability requirements in international bioenergy markets. In: Junginger, M., Goh, C.S., Faaij, A. (Eds.), International Bioenergy Trade, Lecture Notes in Energy. Springer, Netherlands, pp. 125–149.

Port of Rotterdam, 2011. APX-Endex launches the Wood Pellets Exchange. Port of Rotterdam Website.

Porter, M.E., 2008. Competitive Strategy: Techniques for Analyzing Industries and Competitors, first ed. Free Press, New York, NY.

Rodriguez, I., van Dam, J., Reed, A.L., Ugarte, S., 2011. Roadmap to an organized global commodity market of bioenergy carriers. SQ Consult Newsletter.

Roeber, J., 1993. The Evolution of Oil Markets: Trading Instruments and Their Role in Price Formation. Energy and Environmental Programme, Royal Institute of International Affairs, London, UK.

Rost, B., 2015. Interview with Bernhard Rost. Statkraft Financial Energy, Stockholm, Sweden.

Sampson, A., 1975. The Seven Sisters: The Great Oil Companies & the World They Shaped. Viking Adult, New York.

Schernikau, L., 2010. Economics of the International Coal Trade: The Renaissance of Steam Coal. Springer Science & Business Media, Berlin, Germany.

Searcy, E., Hess, J.R., Tumuluru, J., Ovard, L., Muth, D.J., Trømborg, E., et al., 2014. Optimization of biomass transport and logistics. In: Junginger, M., Goh, C.S., Faaij, A. (Eds.), International Bioenergy Trade, Lecture Notes in Energy. Springer, Netherlands, pp. 103–123.

Siliverstovs, B., L'Hégaret, G., Neumann, A., von Hirschhausen, C., 2005. International market integration for natural gas? A cointegration analysis of prices in Europe, North America and Japan. Energ. Econ. 27, 603–615. http://dx.doi.org/10.1016/j.eneco.2005.03.002.

Siqueira, K.B., da Silva, C.A.B., Aguiar, D.R.D., 2008. Viability of introducing milk futures contracts in Brazil: a multiple criteria decision analysis. Agribusiness 24, 491–509. http://dx.doi.org/10.1002/agr.20175.

Talus, K., 2014. EU Energy Law and Policy: A Critical Account. Oxford Scholarship Online, Oxford, UK.

USDA, 2013. Grain Inspection Handbook – Book II: Grain Grading Procedures. US Department of Agriculture, Washington D.C.

Van Vactor, S.A., 2004. Flipping the Switch: The Transformation of Energy Markets (Ph.D.). University of Cambridge.

Vinterbäck, J., Hillring, B., 1995. Förädlade trädbränslen 1995 ("Refined wood fuels 1995") (No. R 1995:28). NUTEK.

Voegele, 2014. Drax considers Developing Additional U.S. Pellet Capacity. Biomass Magazine. Available from: <Biomassmagazine.com>.

Walet, K., 2012. APX–Endex Wood Pellet Exchange Has Cold Start. Maycroft Website.

Walters, M., 2010. Presentation at Argus Biomass 2010. Brussels, Belgium.

Williamson, O., 1979. Transaction-cost economics: the governance of contractual relations. J. Law Econ. 22 (2), 233–261.

Wynn, G., 2011. Analysis: Wood Fuel Poised to Be Next Global Commodity. Reuters. <http://www.reuters.com/article/us-energy-biomass-commodity-idUSTRE74I3NK20110519>.

Yergin, D., 2009. The Prize: The Epic Quest for Oil, Money, & Power. Free Press, New York.

Zaklan, A., Cullmann, A., Neumann, A., von Hirschhausen, C., 2012. The globalization of steam coal markets and the role of logistics: an empirical analysis. Energ. Econ. 34, 105–116. http://dx.doi.org/10.1016/j.eneco.2011.03.001.

CHAPTER 8

Transition Strategies: Resource Mobilization Through Merchandisable Feedstock Intermediates

P. Lamers, E. Searcy and J.R. Hess

Idaho National Laboratory, Idaho Falls, ID, United States

Contents

Abstract

A variety of feedstock types will be needed to grow the bioeconomy. Respective logistics and market structures will be needed to cope with the spatial, temporal, and compositional variability of these feedstocks. At present, pilot-scale cellulosic biorefineries rely on vertically integrated supply systems designed to support traditional agricultural and forestry industries. The vision of the future feedstock supply system is a network of distributed biomass processing centers (depots) and centralized terminals. This introduces methods to increase feedstock volume while decreasing price and quality supply uncertainties. Depots are located close to the resource, while shipping

Developing the Global Bioeconomy.
DOI: http://dx.doi.org/10.1016/B978-0-12-805165-8.00008-2

2016 Published by
Elsevier Inc.

and blending terminals are located in strategic logistical hubs with access to high bulk transportation systems. The system emulates the current grain commodity supply system, which manages crop diversity at the point of harvest and at the storage elevator, allowing subsequent supply system infrastructure to be similar for all resources. The initiation of depot (pilot) operations is seen as a strategic stepping stone to transition to this logistic system. A fundamental part of initiating (pilot-) depot operations is to establish the value proposition to the biomass grower, as biomass becomes available to the market place only through mobilization. A feedstock supply industry independently mobilizing biomass by producing value-add merchandisable intermediates creates a market push that will derisk and accelerate the deployment of bioenergy technologies. Companion markets can help mobilize biomass without biorefineries. That is, depots produce value-added intermediates that are fully fungible in both a companion and the biorefining market. To achieve this, a separation between feedstock supply and conversion industry may be necessary.

8.1 OBJECTIVE AND LINK TO PREVIOUS CHAPTERS

Building on the analyses done in earlier chapters, this chapter details and discusses transition strategies to bridge the current cellulosic biorefinery feedstock supply system, based on low energy and low bulk density, instable formats (eg, bales), to an advanced feedstock supply system, entailing merchandisable intermediates of high energy and bulk density that are stable and flowable (eg, pellets).

In chapter "Biomass Supply and Trade Opportunities of Preprocessed Biomass for Power Generation," Batidzirai et al. show that feedstock supply systems benefit from preprocessing, such as densification, which increases bulk and energy density but also improves flowability, physical homogeneity, and storability. This reduces transport costs and greenhouse gas (GHG) emissions per energy unit, allowing an increased sourcing radius. International sourcing is already established industry practice in today's biopower and "first-generation" biofuel industries. The biofuel industry, relying on oil, starch, and sugar crops, has seamlessly integrated with the existing grain, vegetable oil, and sugar logistics supply chains. The cellulosic biofuel industry however still relies on regional, conventional logistic patterns and market structures (eg, direct, annual contracts with growers). The biopower industry finds itself in between these extremes. While wood pellets fulfill requirements for density, stability, flowability, etc., international sourcing is not yet organized in a commodity market. In fact, as shown by Olsson et al. in chapter "Commoditization of Biomass Markets," an early commodity exchange platform for wood pellets in Rotterdam failed. At this stage, wood pellet supply contracts are directly agreed between buyer (energy utility) and seller (producer).

To reach a competitive minimum fuel selling price (MFSP), cellulosic biorefineries will require economies of scale. The scale-up of current, first-of-a-kind plants will depend on the ability of the logistics system to supply large quantities of homogeneous feedstock intermediates to allow for a continuous operation. The transition towards such a system will need to bridge logistical as well as market structures. In chapter "Commodity-Scale Biomass Trade and Integration With Other Supply Chains," Searcy et al. provide several suggestions on how existing supply chains could be utilized by the feedstock industry. Olsson et al. address market structures in chapter "Commoditization of Biomass Markets." This chapter builds upon these analyses and lays out specific logistical and market stepping stones to bridge the current to a future bioeconomy feedstock supply system.

8.2 CHALLENGES WITHIN LARGE-SCALE BIOREFINERY FEEDSTOCK SUPPLY CHAINS

Concerns about the sustainability of food and fodder crop-based biofuels (see chapter: "Sustainability Considerations for the Future Bioeconomy" for details) have spurred the promotion of advanced biofuels, part of the future bioeconomy, which focus on nonfood or feed biomass resources such as residual biomass from agricultural, forestry, and other industry operations (eg, municipal solid waste); energy crops; or woody biomass. Harvest operations in agriculture and forestry, however, are aimed to maximize the yield of the target crop (eg, grain), not the residual fraction. This implies that residual biomass is highly variable both spatially and temporally (Kenney et al., 2013); which has implications for the operation of large-scale biorefineries that have a relatively constant demand. Some of the critical uncertainty factors in cellulosic biofuel supply chains, influencing the effectiveness of their configuration and coordination in the system, include uncertainty with respect to feedstock supply quantity, quality, and price (influenced by weather, etc.), biorefinery demand fluctuations linked to biofuel and oil price changes, as well as regulatory and policy changes (see also Sharma et al., 2013). An overview of potential challenges is presented in Table 8.1.

Table 8.1 Summary of the challenges associated with biomass supply chain operations

Challenges	Description of issues
Technical and technological	Geospatial and temporal variability of biomass Compositional variability of biomass Biomass quality impacts conversion yields Current supply chains are designed around high-value agriculture and forestry products, not around optimizing residue extraction
Financial	High capital costs (CAPEX) Drive to reduce CAPEX (smaller plants) *vs.* economies of scale (larger plants) Technical complexity limiting access to finance Risks associated with new technologies (insurability, performance, rate of return) Extended market volatilities (energy and food markets)
Social	Potential health and safety risks Pressure on transport sector Public acceptance ("Not-In-My-Backyard")
Environmental	Consistent, international sustainability frameworks and respective governance Safeguarding of best management practices (BMP) Safeguarding GHG benefits across the whole supply chain Safeguarding water and soil conservation
Policy and regulatory	Impact of fossil fuel tax on biomass transport Underestimation of feedstock challenges Heavy focus on solving technological issues, limited attention to market barriers (eg, fossil fuel subsidies)
Institutional and organizational	Varied ownership arrangements and priorities among supply chain parties Lack of feedstock supply cooperatives Limited number of success stories Growers and traders are reluctant to expand their portfolio to include lesser value products (eg, cellulosic biomass) without dedicated, multiple markets to sell to

8.3 FEEDSTOCK SUPPLY SYSTEM TYPES: CONVENTIONAL AND ADVANCED

Feedstock variability with respect to quantities and resulting changes in supply costs (ie, prices) are largely associated with irregular harvest volumes, linked to inclement weather (eg, droughts) and other conditions affecting harvest timing. Variations in quality, particularly for agricultural residues, are linked to natural, compositional and introduced variability. Empirical data for corn stover, for example, suggests that the harvest year has the strongest effect on compositional variation (eg, physiological ash or carbohydrate content), followed by location and plant variety (Templeton et al., 2009). Harvest practices add an additional layer of complexity. For instance, single-pass harvest where combine and baling operation are done at once and the residue does not touch the ground, ash contamination (ie, introduction of soil) is significantly reduced in comparison to conventional, multipass harvest, where a separate combine and baling operation takes place (Hess et al., 2009).

To supply a national or global bioeconomy, logistics and market structures will need to address and cope with the spatial, temporal, and compositional variability of biomass. Only a reduction of this variability, that is, a constant, large quantity supply within quality specifications, can guarantee stable and high conversion yields necessary for a viable business operation such as a cellulosic biorefinery relying on these supply streams.

At present however, pilot-scale cellulosic biofuel production facilities rely on vertically integrated feedstock supply systems designed to support traditional agricultural and forestry industries, hereafter referred to as conventional systems, where feedstock (predominantly agricultural residues such as wheat straw and corn stover) is procured through contracts with local growers, harvested, locally stored, and delivered in low-density format to the nearby conversion facility (Fig. 8.1). These conventional systems were designed to support traditional agricultural and forestry industries. The conventional system has been demonstrated to work in a local supply context within concentrated supply regions (eg, the US Corn Belt).

Different analyses (Hess et al., 2009; Argo et al., 2013; Jacobson et al., 2014a; Muth et al., 2014) suggest that the conventional system may not be able to achieve high-volume, low-cost feedstock supply outside of high biomass yield regions and could even encounter issues in highly productive areas in some years due to inclement weather (eg, drought, flood, heavy moisture during harvest, etc.). High-volume, low-cost feedstock supply, however, is a prerequisite for the advanced biofuel industry to scale-up and become (more) competitive with fossil-fuel-derived alternatives. Furthermore, feedstock supply uncertainties tend to increase the risk, which (to some extent) has limited the cellulosic biorefinery concept

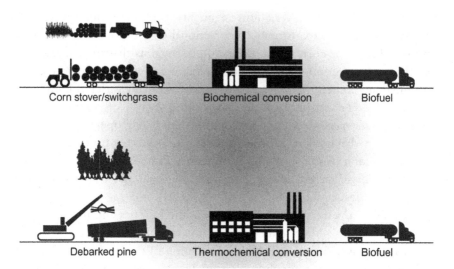

Figure 8.1 Schematic design of select operations in the conventional feedstock supply system.

from being broadly implemented. Advanced feedstock design systems (Hess et al., 2009) introduce methods to reduce feedstock volume, price, and quality supply uncertainties. Advanced systems are based on a network of distributed biomass preprocessing centers (depots) and centralized terminals/elevators (Fig. 8.2). In this system, depots are located close to the biomass resource while shipping and blending terminals are located in strategic logistical hubs with easy access to high bulk transportation systems (eg, rail or barge shipping).

A fundamental difference between the two logistics systems is that the conventional system relies on existing technologies and agri-business systems to supply biomass feedstocks to pioneer biorefineries and requires biorefineries to adapt to the diversity of the feedstock (eg, square or round bales, silage, etc.). The advanced system on the other hand emulates the current grain commodity supply system, which manages crop diversity at the point of harvest and at the storage elevator, allowing subsequent supply system infrastructure to be similar for all resources (Hess et al., 2009; Searcy and Hess, 2010) (Fig. 8.3). Through preprocessing (at the depot) and blending (at the terminal), the variability within the system is reduced significantly in terms of quality and quantity, thus also stabilizing cost/price projections.

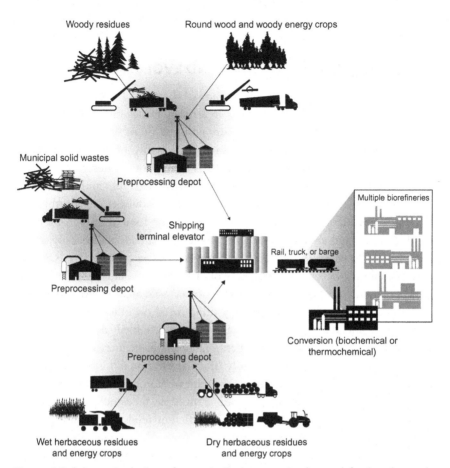

Figure 8.2 Schematic design of a vertically integrated advanced feedstock supply system concept.

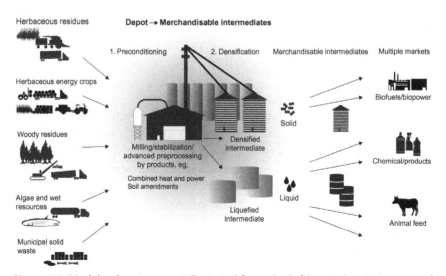

Figure 8.3 Modular depot concept illustrated for multiple biomass input streams and market options for merchandisable feedstock intermediates.

8.4 DEPOT CONFIGURATIONS AND EVOLVEMENT

Common pelleting (ie, densification and stabilization) has enabled the forest industry to trade woody biomass in large volumes internationally. This is a transition from the previously dominating trade of wood chips, having moisture contents of up to 50%, which could only be traded locally or cross-border for low-value markets such as energy, or needed to be of high quality to access distant, higher-value markets such as pulp and paper.

Due to the low-density format of agricultural residues, traditional thinking suggests cellulosic biorefineries are best suited to be located in high-biomass-yielding areas, and should be designed to handle single feedstock of similar format such as wheat straw bales or corn stover (Hess et al., 2009). Regional preprocessing however, through a network of depots, would allow biorefineries to be built almost anywhere, including lower yield areas (Argo et al., 2013). This would not only allow biorefinery siting based on other, often very relevant criteria, including tax incentives, infrastructure, trained labor, etc., but may also prevent potential resource competition among biorefineries.

Individual depots could not only increase energy and bulk density, but also include quality management to achieve compositional homogeneity and specific cost targets by blending multiple feedstock. A network of depots could supply biorefineries with sufficient feedstock (volume), possibly from different biomass in a variety of forms (eg, square and/or round bales, chipped, bundled, raw, etc.). Depots would have a continuum of functionality, from a "standard depot" that would, at a minimum, include particle size reduction, moisture mitigation, and densification, to "quality depots" which may include additional preprocessing steps such as leaching, chemical treatment, or washing (see Lamers et al., 2015a for a detailed techno-economic analysis of different depot configurations).

8.4.1 Early-Stage Depots

The first depots to emerge would likely focus on improving feedstock stability (for storage), increase bulk density (for transport), improve flowability (for stable in-feed rates), and reduce dry matter loss (DML). Influencing feedstock quality is a result of these activities rather than a primary target of the operation. Passive quality management is optionally possible via feedstock blending.

Indirect quality impacts include, for example, drying, which is done to prevent DML. Consistent moisture levels however also benefit conversion efficiency and improve in-feed. Pelleting is done to increase bulk density and transportability, a key aspect in derisking the feedstock supply system. At the same time, using pelleted feedstock also reduces contamination as it sterilizes (through compression and drying). Small-diameter components, including impurities such as soil are drained in the liquor stream of the conversion pretreatment steps (eg, deacetylation).

To address feedstock stability, bulk density, and flowability issues, depot process flow would likely include particle size reduction, moisture mitigation, and densification. An example of an early-stage depot is a common pelleting process involving a two-stage size reduction (grinding), drying, and pelleting. Additional modifications could be made to make the process more efficient. An example of such modifications could include a high-moisture pelleting process; which varies in process sequence, dryer type, and size compared to the common pelleting process.

8.4.2 Later-Stage Depots

As more depots enter the marketplace, they would evolve from focusing on addressing format and creating a uniform product, to actively addressing feedstock quality aspects specific to the end-use market it targets, for example, cellulosic biorefineries, animal feed, or the heat and power sector. They could produce enhanced feedstock (with lower contamination levels) or even process intermediates and thus reduce the pretreatment requirements at the client facility (Jacobson et al., 2014b; Lamers et al., 2015b). To match end-use markets, various kinds of pretreatment steps are possible within these "quality" depots. Thermal pretreatment technologies (eg, torrefaction) create feedstock with structural homogeneity and superior handling, milling, and cofiring properties. Chemical pretreatment changes the composition and structure of the biomass. This reduces the energy required to grind or densify the feedstock, improves flowability and storage stability, and removes contaminants detrimental to downstream biorefinery processes.

An example of addressing quality at the depot is the ammonia fiber expansion (AFEX) process. AFEX is a promising pretreatment that involves an ammonia-based process resulting in physical and chemical alterations to lignocellulosic biomass that improves their susceptibility to enzymatic attack (Bals et al., 2011). As part of a depot concept, AFEX pretreatment of corn stover and switchgrass has been shown to generate a higher return on investment compared to other depot configurations, for example, wood-based pyrolysis facilities (Bals and Dale, 2012). Furthermore, AFEX pellets can be sold to animal feed operations.

8.5 DEPOT DEPLOYMENT

8.5.1 Overcoming the Mobilization Gridlock via Merchandisable Intermediates for Multiple Markets

Feedstock supply systems are currently in a gridlock, where growers will likely not invest in a depot due to slow market growth of biorefineries and the current, limited demand from a single, regional client (biorefinery). On the other hand, biorefineries continue to be limited in expanding their operations in size and number due to high feedstock supply variability in quantity, quality, and price. Thus, a market for feedstock intermediates generated by decentralized depots will not emerge by itself. Rather, a transition strategy is required to break the current development gridlock.

The advanced system is seen as a mature logistical and market structure in which multiple depot types and transloading terminals operate in a high volume (ie, liquid) and competitive feedstock market to serve multiple industries in the bioeconomy. A stepwise introduction of the depot concept is seen as an organic transition towards this vision; yet depots alone do not represent the advanced supply system.

A fundamental part of initiating (pilot-) depot operations is to establish the value proposition to the biomass grower, as the biomass becomes available to the market place only through mobilization. Mobilization is creating the economic drivers required to catalyze the infrastructure investment and biomass resource development investment necessary to transition biomass from available resource, that is, what is on the field, to a merchandisable resource, that is, what is available for sale (Box 8.1).

BOX 8.1 From *Merchandisable* to *Tradable* to *Commodity* Intermediates

Merchandise is defined as goods bought and sold in business, that is, commercial wares. Biomass as such is merchandisable in the form of a good/product that can be sold and bought. Typically, this would imply a minimum form of processing such as cutting, chipping, or baling, and some requirements to storability and transportability. A good becomes tradable when it can be sold in a different location than production. This is practiced, for example, by the wholesale and retail business. At this stage, the product (ie, biomass) will need to comply with more strict requirements to storability and transportability. As a commodity, the good will adhere to a standard quality, which enables physical interchangeability and pricing. A commodity market is highly liquid (in terms of quantity/volume) and competitive (ie, numerous suppliers and buyers). Thus, it is often a supraregional market (see chapter: "Commoditization of Biomass Markets" for details).

The current paradigm for developing feedstock supply systems is that it requires a market pull (ie, new biorefineries) to mobilize the resources. A feedstock supply industry that would independently mobilize biomass by producing value-added merchandisable intermediates however creates a market push that will derisk and accelerate deployment of bioenergy technologies. Accomplishing this would still require a market pull, but initially the pull comes from existing markets; thus, the need for multiple markets (Fig. 8.3).

An obvious question emerges: how do you mobilize biomass into the marketplace without biorefineries to purchase the feedstock? The answer: with companion markets. That is, depots that produce value-added product intermediates that are fully fungible into both the companion market and the biofuels refining market. The stronger, established companion market mobilizes the biomass resource, and that mobilization pushes the second-generation biofuels market into existence. Examples of such markets are biopower or animal feed operations (Fig. 8.3).

Biopower, including cofiring coal with biomass, provides an opportunity to demonstrate value-added advanced preprocessing, developing technologies and systems that could be leveraged by a growing biofuels industry. An example of a technology under development that would minimize, if not eliminate, the need for retrofitting existing coal plants is thermal treatment, such as torrefaction. Torrefaction, and advanced preprocessing technology that could be integrated into a depot slowly dries the biomass to remove essentially all water, such that the feedstock is very friable. The hope is that torrefied feedstock could be handled, fed, and burned very much like coal (Nunes et al., 2014). Another example is AFEX pretreatment (Bals and Dale, 2012), which is value-added processing of biomass for biochemical conversion, but is also a value-added intermediate product for livestock feed. Yet another example is applying thermal treatment strategies to produce oil product with acceptable "shelf-life" as a value-added intermediate for oil-refining routes, but also produces value-added products for heat, power, and specialty markets (liquid smoke, cosmetics, etc.). Technologies are being developed to support the production of a stable bio-oil intermediate at depot-scale.

8.5.2 Separating the Vertical Supply Chain

Continuing along the vertically integrated path that pioneer cellulosic biorefineries have taken will constrain the bioenergy industry to very high-biomass-yielding areas, limiting the industry's ability to increase in

---

BOX 8.2 Pricing and Markets

Contract pricing is probably the most common strategy for longer-term supply agreements. The negotiated price may be connected to an escalation (or de-escalation) clause to allow an adaptation over time. In a cash or spot market, commodities are traded for immediate delivery based on current prices. Companies typically try to avoid relying too heavily on spot markets as it leaves them vulnerable to price fluctuations. In a futures market, delivery is due at a later date and exchange terms are based on a forward price. To offset future market risk/uncertainty, traders apply hedging strategies. An example of a hedge would be if the owner of a stock sold a futures contract stating that he/she will sell the stock at a set price; therefore avoiding market fluctuations.

---

scale to larger plants, and scale as an industry at any size plants. Advanced feedstock supply systems, and depots in particular, enable a bifurcated profitable feedstock supply industry, independently viable from the biofuels industry.

To advance the cellulosic biofuels industry, a separation between feedstock supply and conversion is necessary. Thus, in contrast to the vertically integrated supply chain with a single industry, there are two industries in the advanced feedstock supply system: a feedstock industry and a conversion industry. The split is more beneficial for feedstock producers as they are able to sell into multiple markets. Unless there is a heavy competition among feedstock producers, the split may actually be disadvantageous for the conversion industry; which currently enjoys a monopsony situation (in which it is the sole demand party). Therefore, it may come to a hybrid system in which some upstream activities by the biorefineries secure a bulk of the feedstock and some volumes are bought via contracts (or the spot market) (Box 8.2).

8.6 MARKET TRANSITION

8.6.1 Transition Periods

The depot concept is applied in the forest and agricultural sector, such as wood pellet production and the food/fodder industry (eg, farmer cooperatives). It is not yet widely used however with respect to agricultural residues, and is not part of current biorefinery feedstock supply chains. The appearance of depots across the biorefinery supply chain is expected to occur organically as biofuel producers add preprocessing equipment and storage to existing infrastructure to primarily buffer supply quantity and price risks.

Jacobson et al. (2014) provide a conceptual cost evaluation of this concept as part of an exploration of different feedstock supply management strategies via systems dynamic modeling. In a series of simulations, the authors show the resilience of the concept under several perturbation scenarios, such as weather-related regional supply shortages. The configuration does entail higher cost to the biorefinery but proved to handle volume and price risks better than without storage options. Thus, the investment in small-scale pelleting for short-term volume buffering provides long-term benefits to the biorefinery and becomes a critical element in derisking the supply system.

The organizational structure of a depot can be independent of the biorefinery (Table 8.2). While the biorefinery may own one or several depots, the depots could also be owned and operated by farmer cooperatives, in line with the historical growth of the US grain elevators. Depots may have various business models, that is, they could operate independent from the biorefinery, be owned and operated by the biorefinery, or even be owned by the biorefinery but operated by the producer, consistent, for example, with trends of US grain elevators. Depot location will most likely be driven by the ownership profile (biorefinery vs cooperation) as well as the existing logistical infrastructure (eg, rail lines, shipping terminals) and it is more likely that as quality becomes more uniform the further away the depot will be located from the biorefinery. This makes decentralized locations possible, also in low-yield areas.

The location decision for a depot is driven by the feedstock supply and existing logistical infrastructure (eg, rail lines, shipping terminals) and other factors (eg, socioeconomic). It becomes more independent from the biorefinery location as the improved feedstock material can be transported over long distances with minimal additional costs. This decentralizes biorefinery locations and also incorporates biomass from low-yield areas that would currently be too costly for a conventional supply system.

Table 8.3 provides a suggestion for potential transition periods of the industry along several key depot characteristics. The main influencing factors include ownership structures, as well as location and sizing decisions, which relate to specialized (single feedstock) or flexible (multifeedstock) depots. The type, number, and size of the end-product market will significantly drive the deployment speed of the depot concept.

Table 8.2 Market challenges and opportunities of the depot concept

Challenges/ opportunities	Explanation
Ownership	The ownership and organizational structures behind a depot directly influence the business strategy/behavior (including contractual issues between a depot and biorefineries, etc.).
Sizing and location	The initial depot concept entails distributed entities located in proximity to the biomass source; potentially followed by connections to terminals where feedstock is consolidated prior to further (bulk) distribution. Depot size will be defined by the sourcing radius and the respective biomass availability (year-round). Depot size influences economies of scale. The resulting question is whether optimal depot sizes exist and to what extent economies of scale can be utilized.
Single versus multifeedstock	Feedstock availability/seasonality will influence the depot size and technical layout. To be operating all year, feedstock flexibility will be key. Most likely depots will rely on field-storage options as the conventional system.
Single versus multiproduct	The flexibility of the depot to supply products to multiple markets will be crucial in defining its operational business risks and thus the attractiveness of an investment. A conventional pelleting operation could target industrial as well as residential heat markets (ie, produce to match different technical standards). The AFEX depot already produces pellets that can be applied in the biochemical conversion as well as the feed/fodder industry.
Preprocessing intensity	The level of preprocessing intensity at the depot depends on a number of factors including the typical markets it will sell to, size (economies of scale), biomass availability, access to capital, business strategy, etc.
Waste streams and treatment	Depending on the involved technical processes, depots may generate waste streams effluents that require treatment. The economic viability of a waste-water-treatment facility at a depot directly relates to depot size and profit. Thus, it appears that only larger, highly specialized depots would be able to compensate for such an investment.
	Depots as well as biorefinery feedstock reception stations will create solid waste (eg, broken bales). Creating this organic material closer to the field creates options for reuse and reduction of transport costs to do so. Also, it may serve as a second income stream for the depot.

Table 8.3 Depot transition periods by elements of characteristics

	Short term	Medium term	Long term
Ownership	Biorefineries (upstream investment)	+[a] Farmer cooperatives	+ Third-party (proven business model)
Location	At the biorefinery	+ High-yield areas, centralized location	+ Decentralized, low-yield, stranded feedstock areas with conglomeration in terminals
Single *vs.* multi feedstock use	Single feedstock	+ Specialized depots, high-yield producing regions	+ Multiple feedstock, blending option
Sizing	Pilot and small-scale (<35,000 t/year)	+ Medium to large-scale (>70,000 t/year)	
Preprocessing intensity	Conventional pelleting	+ Advanced (multiple markets)	
End-use markets	Biorefineries/ biofuels	Multiple national/ regional markets, for example, cattle feed, biorefineries	Multiple, potentially international markets

[a]'+' indicates additions to the previous, earlier transition periods.

8.6.2 Supply Chain Opportunities

In a direct comparison, the conventional system actually has lower supply costs than an advanced system within limited (typically 80 km) radius around the biorefinery when it is situated in a high-biomass-yielding region, such as corn stover in Iowa. However, with an increasing sourcing radius and moving outside of high-yielding regions, supply costs in the conventional system increase exponentially while they remain almost stable in the advanced feedstock supply system (Muth et al., 2014). This allows the biorefinery to tap into resources that are not in its immediate vicinity and thus leverage supply risks in years with reduced harvests, for example, due to droughts. Fig. 8.4 shows drought effects across a sourcing

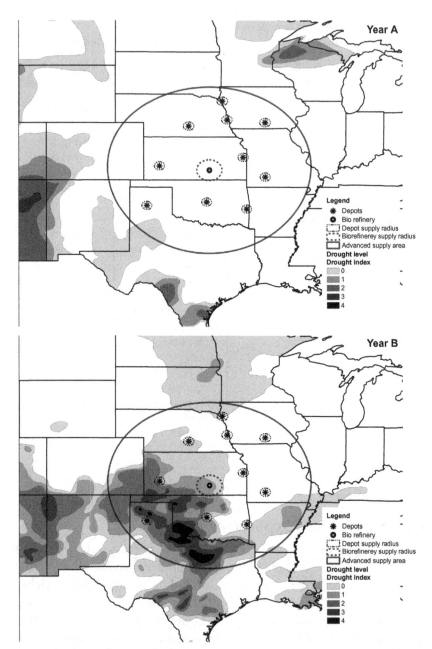

Figure 8.4 Biorefinery and depot locations within a sourcing area (27 miles, dotted line; 500 miles wider circle) in two different years with varying drought patterns (Hartley, 2015).

area of a biorefinery located in the southern part of Kansas. It illustrates that without depots that have access to feedstock outside the heavily affected area, the biorefinery would run a heavy risk of not being able to source sufficient material (within reasonable cost limits) to allow a continuous operation.

Advanced systems address many of the feedstock supply risks associated with conventional systems, and create wider system benefits; most of which translate into cost benefits and risk reduction across the entire biofuel supply chain. Cost benefits of an advanced feedstock supply system include, among others, supply risk reductions (leading to lower interest rates on loans), economies of scale, conversion efficiency improvements, and reduced equipment and operational costs at the biorefinery. Table 8.4 summarizes these benefits and compares them in US$ per liter of gasoline equivalent (LGE) produced. The reason for this measure is the US Department of Energy's advanced biofuel production cost target of $0.79 per LGE by 2022. Note that this list of potential benefits is not exhaustive.

As can be observed from Table 8.4, the cost of additional infrastructure in advanced systems (such as the depot) is more than offset by the savings that advanced feedstock supply systems enable across the entire biofuel supply system. Advanced systems, when matched with the appropriate mode of transportation, could help reduce temporal and spatial biomass variability and allow access to greater quantities of sustainable biomass within a cost target by decoupling the biorefinery from feedstock location. Additionally, densified feedstock has better flow characteristics, which improves transport to and within the biorefinery (via rail, ship, conveyor belts, etc.). This extends the sourcing radius for the biorefinery well beyond the typical 80-km radius of a conventional bale supply and mitigates risks associated with feedstock intermissions (eg, due to adverse weather, pests, and resulting competition for the remaining feedstock within close range). The biorefinery should thus be less vulnerable to feedstock volume, quality, and price volatility (affecting its profitability) and may not need to contract directly with feedstock producers. Reducing profitability risks could also help leverage the reluctance from the investment community to invest in larger facilities, enabling production economies of scale. The variability of feedstock supply to biorefineries in terms of both quality and quantity is recognized as an investment risk by financial institutions. Reducing the variability of feedstock supply will reduce associated project risks which will

Table 8.4 Comparison of the additional preprocessing costs for an advanced feedstock supply system and its biorefinery investment and operation cost-benefits (Lamers et al., 2015b)

	Biochemical conversion plus depot processing ($/LGE)	Thermochemical conversion plus depot processing ($/LGE)	Biochemical conversion plus AFEX pretreatment ($/LGE)
Supply system benefits			
Interest rate reduction of −2% to −5%	−$0.05 to −$0.01	−$0.05 to −$0.02	−$0.05 to −$0.01
Economies of scale (>2000DMT/day)	−$0.19 to −$0.13	−$0.17 to −$0.11	−$0.19 to −$0.13
Conversion yield improvements	−$0.14 to −$0.06	−$0.13 to −$0.04	−$0.14 to −$0.06
Reduced storage equipment at the biorefinery	−$0.04	−$0.04	−$0.04
Reduced handling equipment at the biorefinery	−$0.13	−$0.08	−$0.13
Reduced pretreatment equipment at the biorefinery	Not applicable	Not applicable	−$0.11
SUM benefits	−$0.54 to −$0.36	−$0.47 to −$0.27	−$0.65 to −$0.48
Processing/ pretreatment costs at the depot	$0.09 to $0.15	$0.15	$0.19
TOTAL	−$0.45 to −$0.21	−$0.32 to −$0.12	−$0.46 to −$0.29

Note: Not all benefits associated with the depot concept have been quantified and included in this table.

be reflected in the weighted average cost of capital for financing biorefineries. Also, advanced systems will reduce the handling infrastructure (for raw biomass in various formats) at the biorefinery, improve in-feed operations and thus reduce investment and operating costs. This should further reduce investment risks.

8.7 CONCLUSIONS

The cellulosic bioenergy industry is still in its infancy. However, a handful of US biorefineries had to implement feedstock supply strategies for commercial facilities. The vertically integrated feedstock supply systems developed by each of these biorefineries are similar to conventional supply system designs described in this section. Cost estimates for establishing the feedstock supply system on existing biomass resources (ie, existing crops, such as corn stover, rather than energy crops) have ranged from 30% to greater than 50% of the cost of the biorefinery.

Recently published analyses show that biorefineries, cellulosic or otherwise, that are attached to an existing and fully mobilized feedstock resource (corn stover being the reference feedstock) have a reduced risk profile that may translate into a 2–5% financing interest rate reduction (Hansen et al., 2015; Lamers et al., 2015b). Depending on the size of the biorefinery, this could resemble as much as a $0.05 per liter reduction in unit production cost over depreciable facility life. Additionally, evidence also suggests that facilities that do not have to develop their own supply systems can be fully operational 12–18 months sooner than those that must build supply systems. Combining the cost savings and faster start up time, there is great incentive for biorefineries to move beyond conventional supply systems.

The advanced feedstock supply system is envisioned to be a mature logistical and market structure in which multiple depot types and transloading terminals operate in a high volume (ie, liquid) and competitive feedstock market to serve multiple industries in the bioeconomy. A stepwise introduction of the depot concept is seen as an organic transition towards this vision; yet depots alone do not represent the advanced feedstock supply system. Initially, depots could entail solely processes to stabilize biomass for storage and transport. They could be owned by the biorefinery to buffer supply variations and reduce storage footprint and harmonize in-feed operations. Fully independent depots as well as advanced technical designs, for example, dilute acid pretreatment, may only emerge over time.

A critical component within the advanced system is the biomass processing depot. Depots, and more importantly advanced preprocessing technologies hosted at depots, can mitigate many risk factors faced by biorefineries associated with conventional supply systems, such as aerobic instability (ie, rotting and fire risk), high-quality variability, inefficient handling and transportation, supply chain upsets (due to weather, pests, etc.), just to name a few. In addition to helping current biorefineries reduce feedstock supply risk, depots essentially "mobilize" biomass resources into the market place by producing value-added merchandisable biomass intermediates that can be traded and aggregated.

REFERENCES

Argo, A.M., Tan, E.C.D., Inman, D., Langholtz, M.H., Eaton, L.M., Jacobson, J.J., et al., 2013. Investigation of biochemical biorefinery sizing and environmental sustainability impacts for conventional bale system and advanced uniform biomass logistics designs. Biofuel. Bioprod. Bior. 7 (3), 282–302.

Bals, B., Wedding, C., Balan, V., Sendich, E., Dale, B., 2011. Evaluating the impact of ammonia fiber expansion (AFEX) pretreatment conditions on the cost of ethanol production. Bioresource Technol. 102 (2), 1277–1283.

Bals, B.D., Dale, B.E., 2012. Developing a model for assessing biomass processing technologies within a local biomass processing depot. Bioresour. Technol. 106, 161–169.

Hansen, J., Jacobson, J., Roni, M.S. 2015. Quantifying supply risk at a cellulosic biorefinery. Proceedings of the 33rd International Conference of the System Dynamics Society, Massachusetts, USA.

Hartley, D., 2015. US Drought Map. Idaho National Laboratory, Idaho Falls, ID, USA.

Hess, J.R., Kenney, K.L., Ovard, L.P., Searcy, E.M., Wright, C.T., 2009. Commodity-scale production of an infrastructure-compatible bulk solid from herbaceous lignocellulosic biomass Uniform-format Bioenergy feedstock supply system design report series. Idaho National Laboratory, USA.

Jacobson, J.J., Carnohan, S., Ford, A., Beall, A., 2014. Simulating Pelletization Strategies to Reduce the Biomass Supply Risk at America's Biorefineries. Idaho National Laboratory, Idaho Falls, ID, USA, Retrieved from <https://inldigitallibrary.inl.gov/sti/6269324.pdf>, [March 15, 2016].

Jacobson, J., Cafferty, K., Bonner, I., 2014a. A Comparison of the Conventional and Blended Deedstock Design Cases to Demonstrate the Potential of Each Design to Meet the $3/GGE BETO Goal. Idaho National Laboratory, Idaho Falls, ID, USA.

Jacobson, J., Lamers, P., Roni, M., Cafferty, K., Kenney, K., Heath, B., et al., 2014b. Techno-economic Analysis of a Biomass Depot. Idaho National Laboratory, Idaho Falls, ID, USA.

Kenney, K.L., Smith, W.A., Gresham, G.L., Westover, T.L., 2013. Understanding biomass feedstock variability. Biofuels 4, 111–127.

Lamers, P., Roni, M.S., Tumuluru, J.S., Jacobson, J.J., Cafferty, K.G., Hansen, J.K., et al., 2015a. Techno-economic analysis of decentralized biomass processing depots. Bioresour. Technol. 194, 205–213.

Lamers, P., Tan, E.C.D., Searcy, E.M., Scarlata, C., Cafferty, K.G., Jacobson, J.J., 2015b. Strategic supply system design—a holistic evaluation of operational and production cost for a biorefinery supply chain. Biofuel. Bioprod. Bior. 9 (6), 648–660.

Muth, D.J., Langholtz, M.H., Tan, E.C.D., Jacobson, J.J., Schwab, A., Wu, M.M., et al., 2014. Investigation of thermochemical biorefinery sizing and environmental sustainability impacts for conventional supply system and distributed pre-processing supply system designs. Biofuel. Bioprod. Bior. 8, 545–567.

Nunes, L.J.R., Matias, J.C.O., Catalão, J.P.S., 2014. A review on torrefied biomass pellets as a sustainable alternative to coal in power generation. Renew. Sustain. Energ. Rev. 40, 153–160.

Searcy, E., Hess, J.R., 2010. Uniform-Format Feedstock Supply System: A Commodity-Scale Design to Produce an Infrastructure-Compatible Biocrude from Lignocellulosic Biomass. Idaho National Laboratory, Idaho Falls, ID, USA.

Sharma, B., Ingalls, R.G., Jones, C.L., Khanchi, A., 2013. Biomass supply chain design and analysis: basis, overview, modeling, challenges, and future. Renew. Sustain. Energ. Rev. 24, 608–627.

Templeton, D.W., Sluiter, A.D., Hayward, T.K., Hames, B.R., 2009. Assessing corn stover composition and source of variability via NIRS. Cellulose 16 (4), 621–639.

Conclusions

P. Lamers[1], E. Searcy[1], J.R. Hess[1] and H. Stichnothe[2]

[1]Idaho National Laboratory, Idaho Falls, ID, United States
[2]Thünen Institute of Agricultural Technology, Braunschweig, Germany

INTRODUCTION

Global primary energy supply is dominated by fossil fuels, accounting for 81% in 2013 (IEA, 2015). Growing global energy needs, due to rising incomes and population trends, project an increase in the use of fossil fuels over the coming decades (IEA, 2014). Fossil fuel usage however has been identified to be one of the key sources of greenhouse gas (GHG) emissions and thus a major contributing factor to anthropogenic climate change (IPCC, 2014). One of the key challenges in the energy sector for this century will therefore be to decouple a growing energy supply from an increase in GHG emissions. Bioenergy has been regarded as one of the options to address this challenge. It already forms an integral part of many sectors, such as biomass systems for food, fodder, fiber, and forest products. In 2013, bioenergy contributed around 10% to the global primary energy supply (IEA, 2015). While the majority of this volume still accounts for traditional biomass use, a little over a third represents the use of biomass for heat and power generation or its conversion into road transportation fuels in industrialized countries, typically coined "modern" usage (IEA, 2014; IPCC, 2014).

Expert reviews of available scientific literature predict potential primary biomass deployment levels of up to 300 EJ by 2050 (Chum et al., 2011). A recent IEA Bioenergy analysis of five globally significant supply chains, including boreal and temperate forests, agricultural crop residues, biogas, lignocellulosic crops, and cultivated grasslands and pastures in Brazil, has confirmed that feedstocks produced via logistically efficient production systems can be mobilized to make significant contributions to achieving global targets for bioenergy by 2050 (IEA-Bioenergy, 2015).

SCOPE

While modern bioenergy deployment levels are heterogeneous across the globe, in some markets, it has already been part of the national fuel mix for several decades. The specific drivers for its deployment vary from

Developing the Global Bioeconomy.
DOI: http://dx.doi.org/10.1016/B978-0-12-805165-8.00018-5

© 2016 Elsevier Inc.
All rights reserved.

187

country to country and have, due to different policy frameworks, industry structures, and natural circumstances (eg, growing conditions), led to the development of distinct supply chains (eg, feedstock choices) and markets (eg, specific sectors). It is on these markets and lessons learned that this book focuses on.

Despite the vast amount of politically driven strategies, there is still little understanding on how current markets will transition towards a national and essentially global bioeconomy. The transition from an economy based on fossil raw materials to a bioeconomy obtaining its raw materials from renewable biological resources requires concerted efforts by international institutions, national governments, and industry sectors, and prompts for the development of bioeconomy policy strategies.

This joint analysis brings together expertise from three IEA Bioenergy subtasks, namely Task 34 on Pyrolysis, Task 40 on International Trade and Markets, and Task 42 on Biorefineries. The underlying hypothesis of the work is that bioeconomy market developments can benefit from lessons learned and developments observed in modern bioenergy markets. The question is not only how the bioeconomy can be developed, but also how it can be developed sustainably in terms of economic (eg, risk reduction and piggy-backing on existing industry) and environmental concerns (eg, nonfood biomass based).

The strength of bringing three IEA Bioenergy subtasks into this analysis is found in each task's area of expertise. Tasks 34 and 42 identify the types of biorefineries that are expected to be implemented and the types of feedstock that may be used. Task 40 provides complementary work including a historical analysis of the developments of biopower and biofuel markets, integration opportunities into existing supply chains, and the conditions that would need to be created and enhanced to achieve a global biomass trade system supporting a global bioeconomy. It is expected that a future bioeconomy will rely on a series of tradable feedstock intermediates, that is, commodities. Investigating the prerequisites for such a commoditization, and lessons learned by other industries, play a central role in this analysis.

The analysis covers an overview of biorefineries in a global biobased economy, identifies feedstock and conversion pathways, and outlines the status of demonstration plants and underlying economics. It brings together lessons learned and case studies from the biopower and biofuel markets and covers a brief historical description of international bioenergy trade and markets and links these and future developments to biomass preprocessing options. Furthermore, it bridges current to future bioeconomy-related markets by identifying and describing logistical integration opportunities. Several case studies of existing supply chains in the bioenergy markets are analyzed with respect to increased volume and end-use markets.

Within the context of this work, bioeconomy is defined as the economic, environmental, and social activities associated with the production, harvest, transport, preprocessing, conversion, and use of biomass for biopower, bioproducts, and biofuels. As such a bioeconomy refers to the set of economic activities that relate to the invention, development, production, and use of biological products and processes (OECD, 2009). The main industrial sectors likely to be involved in the future bioeconomy are agriculture and forestry, and include their related processing industries (eg, food and feed, pulp and paper, etc.), plus chemicals and materials (Bell et al., 2014; NEA, 2014). While the production of biobased materials is not new, the majority of fuels, nitrogen fertilizer, organic chemicals, and polymers are still derived from fossil-based feedstock, predominantly oil and gas.

LESSONS FROM BIOENERGY

Whether the product is an energy carrier such as biofuel or biopower, or a biomaterial or biochemical, its production requires biomass as a feedstock. Many factors impact the selection of biomass feedstock: the total delivered cost and availability of the biomass, the quality required by the conversion process which may align with one biomass type better than another, as well as the underlying requirement for sustainability. Also, all of these factors are interrelated; a more robust conversion process may be able to use a broader range of feedstocks, increasing biomass availability and potentially decreasing sourcing costs. However, robustness is only one consideration when selecting a conversion process. Additional considerations include (but are not limited to) total capital cost, per-unit production cost, state of technology development of the bioeconomy process, as well as its comparable conventional processes. Also, different conversion processes yield different products, and therefore the desired product (and the markets, existing and anticipated) for these products must be a key consideration. An option that has many advantages is the "integrated biorefinery" approach, accepting a variety of feedstocks and producing a variety of outputs depending on current market and feedstock outlooks.

An opportunity exists to leverage existing infrastructure and technologies, developed and proven by other industries, such as grain and petroleum. Using high-capacity, economic transport and handling equipment developed through these other industries could immediately enhance the efficiency of biomass logistics systems. However, this infrastructure was developed to move dense, flowable, consistent material (whether solid or liquid), characteristics which raw biomass often does not have under current agricultural supply systems. Is it an option to transform raw, unstable, bulky biomass into a flowable commodity like grain or petroleum?

The commoditization of biomass feedstock offers many potential benefits, including access to multiple markets (for both producers to sell into and biorefiners to purchase from), standardization of quality, and price stabilization. Lessons learned from the many other industries that have undergone commoditization (eg, grain and oil), and lessons learned from those which have not (such as coal), can be weighed to establish the best path forward. Raw biomass, currently not a commodity, will require significant investment to transition from the current supply system of the agricultural industry and markets to a commodity-type multiple market system.

Across this book, three themes emerge as fundamental to bridging the gap from the current, bioenergy-focused system to an integrated bioeconomy: value-add, risk mitigation, and performance metric (Table 1).

The value proposition across the supply chain from the biomass grower, to the preprocessing and conversion industry is required to mobilize resources and create market and trade options. Value-add can be achieved in several ways, for example, enhanced market value due to

Table 1 Main themes and their representation across the book

Chapters	Value-add	Risk mitigation	Sustainability
1. Bioeconomy strategies	√		√
2. Second-generation biorefineries	√		
3. Industry status	√		
4. Sustainability considerations			√
5. Preprocessing benefits	√		
6. Integration options		√	
7. Commodity markets		√	
8. Transition strategies	√	√	

√: theme covered in respective chapter.

improved product properties or by more favorable political framework conditions through, for example, carbon pricing. Wood pellets are a key example of value-add through preprocessing, creating lower moisture, higher flowability, higher energy and bulk density intermediates that can access larger markets—geospatially and temporally.

On the implementation side, large-scale biomass investment projects face systematic (wider market/economic) as well as nonsystematic, project-immanent risk. The latter can be mitigated (among others) by plant design and process integration. This book presents market as well as technical integration options, for example, how utilizing existing logistical infrastructure can be leveraged by emerging supply chains. It analyzes factors necessary for commoditization and presents the linkage between merchandisable, tradable, and commodity-type feedstock intermediates.

Sustainability is the key performance metric in the global bioeconomy. Intermediates and final products have to adhere to higher environmental standards than their often fossil-fuel-based substitutes. Eventually, when respective framework conditions level the playing field for products of the global bioeconomy, they will become available to consumers at equal costs, which will drive their demand and deployments.

REFERENCES

Bell, G., Schuck, S., Jungmeier, G., Wellisch, M., Felby, C., Jorgensen, H., et al. 2014. IEA bioenergy Task 42 biorefining: sustainable and synergetic processing of biomass into marketable food & feed ingredients, chemicals, materials and energy (fuels, power, heat). Wageningen, IEA Task 42: 63.

Chum, H., Faaij, A., Moreira, J., Berndes, G., Dhamija, P., Dong, H., et al., 2011. Bioenergy. In: Edenhofer, O., Pichs-Madruga, R., Sokona, Y. (Eds.), IPCC Special Report on Renewable Energy Sources and Climate Change Mitigation Cambridge University Press, Cambridge, UK and New York, USA p. 298.

IEA, 2014. World Energy Outlook. International Energy Agency, Paris.

IEA, 2015. Key World Energy Statistics. International Energy Agency, Paris.

IEA-Bioenergy (2015). Mobilizing Sustainable Bioenergy Supply Chains. Paris, France, Strategic Inter-Task study, commissioned by IEA Bioenergy and carried out with cooperation between IEA Bioenergy Tasks 37, 38, 39, 40, 42, and 43. Available from: <http://www.ieabioenergy.com/wp-content/uploads/2015/11/IEA-Bioenergy-inter-task-project-synthesis-report-mobilizing-sustainable-bioenergy-supply-chains-28ot2015.pdf> (accessed 09.12.15.).

IPCC, 2014. Climate Change 2014: Mitigation of Climate Change. Contribution of Working Group III to the Fifth Assessment Report of the Intergovernmental Panel on Climate Change. Cambridge University Press, Cambridge, United Kingdom and New York, NY, USA.

NEA, 2014. Setting up international biobased commodity trade chains. A guide and 5 examples in Ukraine. Den Hague, the Netherlands, Netherlands Enterprise Agency. Available from: <http://english.rvo.nl/sites/default/files/2014/06/Setting%20up%20 international%20biobased%20commodity%20trade%20chains%20-%20May%202014. pdf> (accessed 04.08.14.).

OECD, 2009. The Bioeconomy to 2030: designing a policy agenda. Paris, France, Organisation for Economic Co-operation and Development: 322. Available from: <http://www.oecd.org/futures/long-termtechnologicalsocietalchallenges/thebio-economyto2030designingapolicyagenda.htm>.

INDEX

Note: Page numbers followed by "*f*" and "*t*" refer to figures and tables, respectively.

Printed in the United States
By Bookmasters